MEMOIRS OF THE SOCIETY FOR ENDOCRINOLOGY
NO. 20

ENDOCRINE FACTORS IN LABOUR

MEMOIRS OF THE SOCIETY FOR ENDOCRINOLOGY

NO. 20

ENDOCRINE FACTORS IN LABOUR

PROCEEDINGS OF A SYMPOSIUM
HELD AT THE UNIVERSITY OF ABERDEEN
ON 19 TO 22 JULY 1972

EDITED ON BEHALF OF
THE SOCIETY FOR ENDOCRINOLOGY BY
ARNOLD KLOPPER AND JEAN GARDNER

CAMBRIDGE

AT THE UNIVERSITY PRESS

1973

Published by the Syndics of the Cambridge University Press
Bentley House, 200 Euston Road, London NW1 2DB
American Branch: 32 East 57th Street, New York, N.Y.10022

Library of Congress Catalogue Card Number: 73–80474

ISBN: 0 521 20158 6

Printed in Great Britain
by Alden & Mowbray Ltd
at the Alden Press, Oxford

CONTENTS

PREFACE

In the matter of scientific conferences we have long since painted ourselves into a corner. Presumably their original purpose was to communicate findings in a more vivid and direct fashion than the written word. That purpose is now frustrated by the growth of subsidiary alternative uses. Some uses are far removed from scientific communication. An offer to read a short paper at a meeting is almost always accepted, if only because the organisers seldom know in advance what you are going to say. As a result conference sessions are sometimes dreary occasions where a series of ten minute papers are gabbled at high speed to a sparse audience of people waiting to give their own paper, while the bulk of the participants disport themselves elsewhere. The pressures have led to the growth of monster conferences, where thousands of participants scurry from one simultaneous session to another, picking up crumbs here and there, but mainly getting more and more bewildered. Not that big conferences are necessarily bad or boring. Indeed large societies are obliged to have large conferences. But they serve a different function.

Having defined what is wrong with present day conferences the solution is easy but expensive. You need a small group; 70–90 people is the largest number from whom you can expect reasonably coherent behaviour, moving as a unit to and fro from teas and meals without too large a tail of stragglers holding up proceedings. Even so with such a number there is already some risk of a few lunatics who hog discussions, ride irrelevant hobby horses, abuse the secretaries or drive the hall porter mad by phoning Milan. Scientists adopt ceremonial postures as readily as aldermen at a banquet, and informality is the only defence against ritual behaviour. We were much helped by the opening speaker, Professor Fuchs, taking off his jacket and addressing the chairman by his christian name. The university setting is a *sine qua non* for this type of conference. Catering, lodgings, meeting rooms and a host of services come at low cost or free. Atmosphere is an ingredient which defies analysis. In the case of the Aberdeen Symposium the 15th century Crown Tower at King's College and the extraordinary appearance of some junior members of staff on the arts side all contributed. Access to university facilities was in large measure due to the fact that the Postgraduate Medical Education Board acted as co-sponsor of the symposium and particular acknowledgement is due to the Postgraduate Dean, Professor Ramsay, for this.

[vii]

The organisational pattern which flowed from this analysis was simple. Invited speakers, preferably from teams who had made a continuous contribution to the endocrinology of labour over some time, were given 40–60 minutes to review their work and place it in the context of knowledge in the field. Then the audience, most of whom were working in the field, had the same length of time to discuss the data and consider the conclusions which might be drawn from them. Although the discussions were a central part of the symposium, their force and spontaneity could not be preserved for publication. Recording discussion is a clumsy and obtrusive device: the highly edited versions of the garbled exchanges seldom have any of the impact of the event. It was suggested that participants could produce a written version of the contribution, but these were so sparse that discussion sections could not be justified. In any case invited contributors were asked to submit a prepared manuscript which corresponded only in sense to what they had said.

Small conferences are uneconomic; desperately so if you have to pay for speakers from the far corners of the earth. We managed to organise in advance the theft of one speaker from Auckland and of another from Sydney who were being subsidised for other meetings. That takes luck and a small speciality like reproductive endocrinology where workers in various parts of the world are peering over each other's shoulders. But most of all it takes money. In this case the money came from pharmaceutical companies. Of course one tries cunningly to imply that there is something in it for them but most medical directors are old hands at the game and our money came from goodness of heart. We are grateful for donations from

Duphar Laboratories Ltd.
Imperial Chemical Industries Ltd.
Messrs May & Baker Ltd.
Organon Laboratories Ltd.
Ortho Pharmaceutical Ltd.
Messrs Parke, Davis & Co.
Sandoz Products Ltd.
Schering Chemicals Ltd.
Messrs E. R. Squibb & Sons Ltd.
Syntex Pharmaceuticals Ltd.
Messrs Upjohn Ltd.
Messrs Wm R. Warner & Co. Ltd.
Messrs John Wyeth & Brother Ltd.

Conferences can do without good weather and get by with poor catering facilities. Without a good secretary they are doomed. Miss Gina Cowie acted as treasurer, secretary and every kind of trouble shooter. That we met at all and survived the experience is in no small measure due to her.

It is likely that oxytocin, prostaglandins, oestrogens and progesterone all play some part in the onset of labour. When the data about each of these were set side by side a new concept of the integrated actions of these hormones began to form. There are still many ifs and buts and the idea is by no means peculiar to the Aberdeen Symposium. But this was one of the first times that it had been clearly formulated in a manner amenable to experimental verification. That is the only reward of the backstage workers in Aberdeen who spent part of their year organising the conference and the only justification for this publication.

ARNOLD KLOPPER,
ABERDEEN

LIST OF CONTRIBUTORS

ANDERSON, ANNE B. M.
Department of Obstetrics and Gynaecology, Welsh National School of Medicine, Heath Park, Cardiff, CF4 4XN. (See title page of article for present address.)

BENGTSSON, LARS P.
Department of Obstetrics and Gynaecology, University Hospital, S-220 05 Lund 5, Sweden.

CHARD, T.
Departments of Chemical Pathology and Obstetrics and Gynaecology, St Bartholomew's Hospital, London, E.C.I.

COX, R. I.
CSIRO, Division of Animal Physiology, Ian Clunies Ross Animal Research Laboratory, P.O. Box 239, Blacktown, NSW 2148, Australia.

CURRIE, W. B.
CSIRO, Division of Animal Physiology, Ian Clunies Ross Animal Research Laboratory, P.O. Box 239, Blacktown, NSW 2148, Australia.

DAWES, G. S.
The Nuffield Institute for Medical Research, University of Oxford, Osler Road, Headington, Oxford, OX3 9DS.

FUCHS, ANNA-RIITTA
The Population Council, The Rockefeller University, York Avenue & 66th Street, New York, N.Y. 10021, USA.

FUCHS, FRITZ
Department of Obstetrics and Gynaecology, Cornell Medical Center, The New York Hospital, 525 East 68th Street, New York, N.Y. 10021, USA.

GILLESPIE, ARNOLD
Department of Obstetrics and Gynaecology, University of Adelaide, Queen Victoria Hospital, Rose Park, Adelaide, South Australia.

KLOPPER, ARNOLD
Department of Obstetrics and Gynaecology, Clinical Research Unit, Maternity Hospital, Foresterhill, Aberdeen, AB9 2ZA.

LIGGINS, G. C.
Postgraduate School of Obstetrics and Gynaecology, National Women's Hospital, Claude Road, Auckland 3, New Zealand.

THORBURN, G. D.
CSIRO, Division of Animal Physiology, Ian Clunies Ross Animal Research Laboratory, P.O. Box 239, Blacktown, NSW 2148, Australia.

TURNBULL, A. C.
Department of Obstetrics and Gynaecology, The Welsh National School of Medicine, Heath Park, Cardiff, CF4 4XN. (See title page of article for present address.)

WONG, M. S. F.
CSIRO, Division of Animal Physiology, Ian Clunies Ross Animal Research Laboratory, P.O. Box 239, Blacktown, NSW 2148, Australia.

INITIATION OF LABOUR –
FACTS AND FANCIES

By FRITZ FUCHS

INTRODUCTION

Why does the mechanism of labour remain an enigma when most other physiological processes have been clarified? Is it because the obstetrician is unaccustomed to physiological observation and deduction, and because the physiologist stops to consider the process of parturition only when he is about to become a father? The purpose of this symposium is to assess current knowledge, to discuss the pros and cons of various hypotheses and, hopefully, to inspire the participants to studies which eventually will lead to a better understanding of the process of parturition. Not until we understand the process fully, can we expect to provide optimal care for mother and child.

That we have travelled from many parts of the world to gather in this distant corner of the British Isles is no coincidence. British Obstetrics has been born, nursed, and brought up in Scotland and the transition of the Art of Midwifery to the Science of Obstetrics, a metamorphosis yet to be completed, began early in this part of the world. This department of the venerable Aberdeen University, under the able leadership of Sir Dugald Baird has made major contributions to the clarification of the influences of social and constitutional as well as endocrine factors on the outcome of pregnancy.

Any attempt to review the hormonal factors in human parturition must stress the unknown facts as much as those which are known. Given the honour of being the opening speaker, I plan to discuss a number of endocrine factors and their possible roles in parturition. I intend to be more provocative than comprehensive, for we shall have the opportunity to focus on each of the factors and discuss them in detail during the symposium.

If the myometrium were completely at rest for the duration of pregnancy and then suddenly developed contractile activity, one could visualise a biochemical process which suddenly permitted the activation of the contractile mechanism. But the uterus is not completely at rest throughout

Supported in part by Grant No. 67-455 from The Ford Foundation.

gestation; it exhibits sporadic contractions with increasing frequency from mid-pregnancy to term, and the uterus has the capacity to contract at any time during pregnancy as shown by its behaviour during the surgical interruption of pregnancy. At term, the transition from the state of pregnancy (with sporadic contractions) to the state of labour (with frequent and rhythmic contractions) often eludes both the patient and the obstetrician because it is so gradual that it is difficult to define when one state ends and the other begins.

The gradual transition from pregnancy to labour supports the assumption that the activity of the myometrium depends on the relationship between factors which are designed to keep the uterus at rest and factors which facilitate contractions. The balance between these opposing forces is obviously delicate because although the average length of gestation is constant for each species, considerable variations do occur. In women, the difference between 'normal' variations and those classified as pathological, such as premature and postmature labour, are based on arbitrary time limits and not on differences in the physiological behaviour of the uterus.

The great variation in length of gestation between species must be genetically determined, but how the gene expression is translated into the complicated mechanism of parturition is not known. Since the length of the menstrual cycle must be genetically determined, the possibility that the length of the gestational cycle also depends on genetic factors in the maternal organism cannot be excluded. However, in those instances in which genetic defects are known to cause deviations from normal gestation length, the genetic defect has always been found in the foetus. The best known example is in Holstein cattle for which the studies of Holm and collaborators (Holm, 1967) have shown that failure to deliver at the expected time is due to a defect in the foetal pituitary gland. Similar syndromes are unknown in man with the exception of the pregnancy with an anencephalic foetus, where prolonged pregnancy is frequent but not invariable. Nevertheless, this gross malformation of the foetus, which also causes endocrine dysfunction, does point to a foetal role in the initiation of labour.

Let us consider the hormonal status of mother and foetus at the onset of labour and discuss how each hormone might be involved in the activation of the myometrium during labour.

THE ROLE OF OXYTOCIN IN LABOUR

The notion that oxytocin is the physiological activator of the uterus in labour is based mainly upon two facts: that the uterus at term is extremely sensitive to oxytocin, and that oxytocin-induced labour cannot be distinguished from spontaneous labour by the pattern of contractions in any other way. Nevertheless, positive proof of the correctness of the assumption has been extremely elusive.

Is it in fact necessary to postulate an activating mechanism in labour? Could not the intrinsic contractile properties of the myometrium be released by the disappearance of an inhibiting mechanism? After all, if a strip of human myometrium is suspended in a muscle bath, it will usually begin to contract rhythmically without apparent stimulation. Such a mechanism is certainly conceivable, but there is considerable evidence against so-called 'spontaneous contractions' of the human uterus *in vivo*.

During the menstrual cycle the human uterus is constantly active, although the pattern of contractions varies with the phases of the cycle. During menstruation, the contractions are very strong, i.e. the amplitude is high. The frequency is higher than during labour, but lower than in mid-cycle. The period at the resting level between two contractions is very short; the next contraction starts almost immediately after the resting level has been reached by the descending leg of the contraction wave. The contraction pattern has been called 'labour-like', but actually it is only labour-like with regard to the amplitude and to the uniformity of the individual contractions. As the cycle progresses, the amplitude becomes lower, but the frequency increases; at mid-cycle the pattern consists of very fast contractions with minimal amplitude. Toward the end of the cycle, the pattern is reversed and a 'menstrual' pattern is usually established before any menstrual bleeding can be observed.

The uterine contractions during the menstrual cycle are often, but in our opinion erroneously, described as 'spontaneous' uterine activity. This activity can be inhibited by agents without any direct effect on the myometrium. It was shown first by Fuchs et al. (1968) that ethanol inhibits uterine activity at blood concentrations which do not significantly affect myometrial strips *in vitro*. Later, Fuchs & Coutinho (1971) showed that expansion of the plasma volume by various isotonic solutions also inhibits the uterine contractions during menstruation. Ethanol and plasma volume expanders both inhibit the release of vasopressin which is a strong oxytocic agent during the menstrual cycle. Vasopressin is mainly concerned with regulation of the water balance of the body and is usually released from the posterior pituitary gland at a low, tonic level, upon which is superimposed the release of greater amounts in response to specific stimuli, one of which is a fall in the extracellular water concentration.

To assume that a reflex mechanism similar to the milk-ejection reflex initiates labour is an attractive hypothesis. Just as the milk-ejection reflex is initiated by the newborn sucking the nipple, one could visualise a 'child-ejection reflex' initiated by a stimulus originating from the foetus which activates, perhaps through afferent nervous pathways, the maternal hypothalamo-posterior pituitary system with the release of oxytocin (and perhaps vasopressin) as the result. The fact that ethanol can inhibit uterine contractions in early labour without influencing myometrial reactivity to exogenous oxytocin supports such an assumption, but indirectly and not excluding other possibilities. The demonstration of oxytocin in the maternal plasma during labour, including the initial phase, would provide more convincing evidence.

Many attempts have been made to demonstrate oxytocin in the maternal blood during labour. Using the sensitive bioassay developed by Fitzpatrick & Walmsley (1965), Caldeyro-Barcia and his group in Montevideo (Coch et al. 1965) were able to demonstrate considerable oxytocin-like activity in jugular venous blood during the second stage of labour. In the first stage, and in peripheral venous blood during the second stage, the levels were close to the sensitivity level of the assay. Although these authors have since improved the sensitivity of the assay, no further clinical studies appear to have been published.

With the advent of immunoassays of protein and polypeptide hormones, several attempts have been made to produce antibodies against oxytocin for use in radioimmunoassays. Specificity is a crucial problem for such assays and not all of the methods described have solved this problem adequately. The most extensive studies have been published in a series of papers by Chard and coworkers. In one of the first papers, Chard et al. (1970) claimed complete specificity and a sensitivity of 0·75 μu./ml plasma. With this method no oxytocin could be demonstrated in the maternal blood during pregnancy or during various stages of labour. In the cord blood, on the other hand, measurable amounts of oxytocin were found in 40 per cent of the samples examined.

On the basis of their initial findings, Chard, Boyd, Edwards & Hudson (1971) considered it possible that oxytocin of foetal origin could be responsible for the initiation of labour. This assumption received further support when it was found that not only cord blood collected after vaginal delivery but also umbilical blood taken at Caesarean sections, with and without previous uterine contractions, contained oxytocin in a certain percentage of cases.

The fact that ethanol can inhibit premature labour as well as spontaneous labour at term (Fuchs, Fuchs, Poblete & Risk, 1967) does not disprove the hypothesis that foetal oxytocin might be the uterine activator in labour. Ethanol given to the mother reaches the foetal circulation rapidly, and presumably exerts the same blocking effect on the release of the foetal posterior pituitary hormones as on the release of maternal oxytocin and vasopressin. However, it is difficult to see how foetal oxytocin could reach the myometrium without being inactivated by the tissue oxytocinase which is found in abundance in the placenta (Branda & Ferrier, 1971).

For a while maternal oxytocin as the activating agent in labour seemed to be disregarded, but in later publications, Chard and coworkers (Gillespie, Brummer & Chard, 1972) reported that oxytocin can be found in maternal blood during labour. In a number of samples collected at various stages of a single contraction–relaxation cycle, one or two contained measurable levels of oxytocin while the others did not. No relation to the stage of contraction was apparent from such multiple sampling. In any event maternal blood contains oxytocin during labour, albeit off and on in a curious fashion, and several explanations are possible. In a recent review, Chard (1972) has summarised the possibilities. It should be pointed out, though, that in Chard's studies only peripheral plasma was used, whereas in animal studies and the studies of the Montevideo group in human subjects mainly jugular venous blood was examined.

In a very recent communication, Kumaresan, Anandarangam & Vasicka (1972) have reported the results of studies using a radioimmunoassay for oxytocin with a sensitivity of 0·25 μu./ml. The assays were performed on whole plasma in a 1:5 dilution. Oxytocin was detectable in the maternal plasma from the fourth week of pregnancy with mean values rising during gestation from 66 to 163 μu./ml. During active labour oxytocin levels increased to $181 \pm 10\ \mu$u./ml but the levels fluctuated rapidly. After delivery, the levels dropped abruptly, but were maintained up to 15 μu./ml as late as six weeks after delivery (Fig. 1).

In another recent paper, Bashore (1972) has described his radioimmunoassay, which has since been further improved. He found oxytocin in the plasma of 13 patients from whom serial samples were taken during labour; in ten the levels increased with cervical dilatation, in the remainder oxytocin levels were either irregular or constant. Most values were below 20 μu./ml plasma, except at delivery when they were higher; two patients had concentrations as high as 90 μu./ml at delivery. Cord blood was examined in seven patients: no oxytocin was found in four, while three had values of 17 to 20 μu./ml.

To increase the confusion, Vorherr (1972), using a highly sensitive bioassay, has failed to demonstrate any oxytocic activity in maternal blood during labour and dismisses the notion of maternal oxytocin playing any role in human labour.

How is one to reconcile these disparate findings and interpretations?
The available studies have not convinced us that the hypothesis of a
'child-ejection reflex' with release of oxytocin from the maternal pituitary
into the blood as the efferent part of the reflex is incorrect. Further studies
are necessary to examine the following points:

(1) Is oxytocin released into the maternal circulation *before* the onset of
labour in order first to sensitise the myometrium and then to initiate
labour?

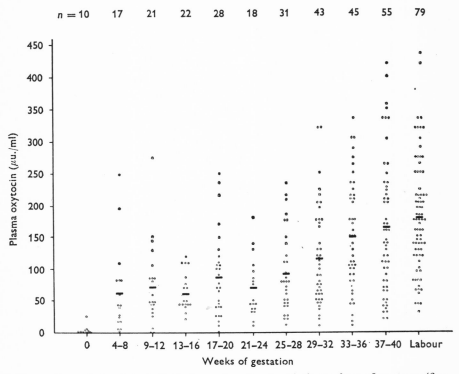

Fig. 1. Oxytocin levels in individual samples of maternal plasma from 285 women (five
of whom were in labour) at various stages of pregnancy and labour, determined by radioimmunoassay. Reprinted with permission from Kumaresan *et al.* (1972).

(2) Is it possible that once labour has been initiated, the myometrium
has acquired such a sensitivity to oxytocin by the interaction of oxytocin
and other factors, that only exceedingly small amounts of oxytocin are
necessary to maintain labour? When labour is induced prematurely with
infusions of oxytocin, the sensitivity of the uterus increases with time;
often several days of infusion are required unless rupture of the membranes
is performed. Amniotomy adds a new sensitising factor, the mechanism
of which is completely unknown.

(3) Could release of oxytocin in spurts of very brief duration provide the myometrium, through binding to the cells, with adequate concentrations while eluding detection in plasma except by extremely frequent sampling? This would provide the best explanation for the findings of Chard and coworkers as well as for the entirely negative findings of Vorherr.

(4) Does oxytocin circulate partially bound to proteins, such as neurophysin, and can this explain the discrepancy between the studies quoted? Could such binding reduce the amounts of oxytocin which can be extracted from the maternal plasma by Chard's method while permitting Kumaresan to get higher values from direct assay of diluted serum. Protein binding is an essential feature of the transport of some other hormones in the circulation. It may be surmised that a similar mechanism plays a part in the transport of oxytocin.

(5) The finding of oxytocin in cord blood at Caesarean section in patients who were not in labour raises the question of when oxytocin first appears in the foetal circulation. Examination of foetal urine, voided immediately after birth by elective Caesarean section, together with an examination of the amniotic fluid, may throw some light on this question.

VASOPRESSIN

As already mentioned, vasopressin may well be the uterine activator during the menstrual cycle, particularly during menstruation. During pregnancy, the human myometrium is less sensitive to vasopressin than to oxytocin and while the sensitivity increases with advancing gestation, the increments are much more moderate than those for oxytocin. Nevertheless, some water retention is usual towards term, while during the post-partum period urine output usually exceeds fluid intake. This could indicate increased release of vasopressin at term and during labour, although Hoppenstein, Miltenberger & Moran (1968), who found considerable amounts of vasopressin in the cord blood, could only detect very low levels in the maternal blood during labour. Vorherr (1972) found $6 \cdot 9 \pm 1 \cdot 7$ μu./ml in the maternal blood during the second stage of labour, and $12 \cdot 8 \pm 3 \cdot 9$, $71 \cdot 0 \pm 7 \cdot 8$ and $65 \cdot 6 \pm 12 \cdot 7$ μu./ml maternal urine during the first, second and third stages, respectively. It seems unlikely that the amount of vasopressin released during labour could be the sole activator of the myometrium but a potentiation of the effect of other possible activators, particularly oxytocin, cannot be excluded. Further investigations of the physiological role of vasopressin in mother and foetus during labour are clearly needed.

CATECHOLAMINES

The human myometrium contains both α- and β-adrenergic receptors. According to the hypothesis of Ahlquist (1948), stimulation of the α-receptors results in the activation of smooth muscle while stimulation of β-receptors results in inhibition of smooth muscle activity. In the myometrium, noradrenaline is mainly α-adrenergic, stimulating contractions, while adrenaline is predominantly β-adrenergic, inhibiting uterine activity. Sjöberg (1967) and Owman, Rosengren & Sjöberg (1967) have used the specific histochemical fluorescence method of Falck (1962) to study the adrenergic innervation of the uterus in various species including man. They found strong evidence that the adrenergic transmitter in the uterus is noradrenaline. They found a particularly rich supply of adrenergic nerves in the cervix. Another finding of interest was the demonstration of sympathetic ganglia in the utero-vaginal junction both in man and in animals. These ganglia give rise to 'short' adrenergic neurones in contrast to the ordinary 'long' adrenergic neurones from pre- and para-vertebral ganglia. Towards the end of pregnancy there was a marked decrease in uterine noradrenaline, and only a few adrenergic nerve terminals persisted. These findings, and particularly the differences between corpus and cervix and between the non-pregnant and pregnant myometrium, may have some functional significance.

The findings of the Swedish group have been extended by others such as Wood and his collaborators in Australia (Wansbrough, Nakanishi & Wood, 1968; Nakanishi, McLean, Wood & Burnstock, 1969) who confirmed the histochemical findings in the human uterus, and studied the action of sympathomimetic and sympathetic blocking agents on uterine strips and the effect of transmural stimulation of intramural nerves. These pharmacological studies also showed the presence of noradrenergic innervation of the myometrium. They showed that the uterine muscle contains both α-excitatory and β-inhibitory adrenergic receptors and that catecholamines exert a greater inhibitory influence on the pregnant than on the non-pregnant uterus.

The most carefully controlled study of the urinary excretion of adrenaline and noradrenaline at the end of pregnancy and after delivery has been carried out by Zuspan (1970). Environmental factors, activity and sleep patterns must be comparable for such a study to be meaningful. Figure 2 shows that the values for adrenaline obtained before and after delivery do not differ from those obtained for the non-pregnant controls. There is a trend towards increased excretion of adrenaline in the 24 h after delivery; however, it is not statistically significant. However, the excretion of noradrenaline in the 24 h period after delivery is two to three times higher

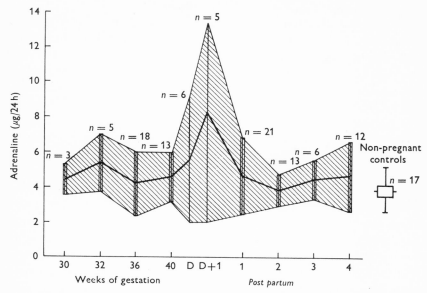

Fig. 2. Excretion of adrenaline in 24-h urine samples from women at various stages of pregnancy, during delivery (D), the 24 h after delivery (D+1), and the first 4 weeks *post partum*. Vertical lines indicate collection of three successive 24-h urine specimens. The heavy middle line shows the means. Reprinted with permission from Zuspan (1970).

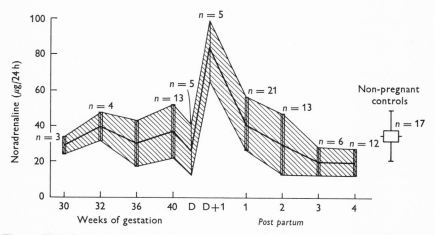

Fig. 3. This figure is schematically identical with Fig. 2 and represents the urinary excretion of noradrenaline. The only significant value is at D+1 (24 h *post partum*). Reprinted with permission from Zuspan (1970).

than it was before labour or later in the period *post partum*, indicating that
noradrenaline release is indeed increased in association with labour
(Fig. 3).

But does this association mean that labour is caused by an activation of
the myometrium by noradrenaline, or is the increased release and excretion
a consequence of the emotional and physical activity of labour? Previous
studies have shown that early labour is not associated with alterations of
adrenaline and noradrenaline concentrations, and the high urinary
excretion in the 24 h after delivery may be no more than a spill over from
the increased strain of late labour.

A large number of pharmacological studies of the effects of the catecholamines and α- and β-adrenergic stimulating and blocking agents have
been carried out in the last few years but interpretation of the results is
hindered by the fact that the effects *in vivo* and *in vitro* are often at variance.
While these studies may provide us with compounds which can inhibit
uterine activity and prevent premature delivery, it is questionable whether
they will clarify the physiological role of the two adrenergic neurohormones
in human labour, and I shall therefore refrain from an attempt to sum-
marise them.

THE PHYSIOLOGICAL ROLE OF THE
PROSTAGLANDINS

The prostaglandins have recently had a meteoric rise on the physiological
horizon. Will they develop into bright and permanent stars or will they
drop out of the picture again like other meteors?

Prostaglandins are usually characterised as tissue hormones and they
seem to be produced in a variety of tissues in response to a variety of
stimuli. The range of effects is extremely wide, but one well-documented
effect is that on smooth muscle, including the myometrium. Thanks largely
to the chemical studies carried out in Sweden, a number of naturally
occurring prostaglandins and many metabolites and analogues are now
known. The prostaglandins of the E and the F series have aroused most
speculation in the field of obstetrics, particularly prostaglandin E_2 (PGE_2)
and prostaglandin $F_{2\alpha}$ ($PGF_{2\alpha}$). These two and two other closely related
prostaglandins have been demonstrated in the amniotic fluid, and $PGF_{2\alpha}$
has been found in the maternal circulation during labour and abortion.

Karim, Trussell, Patel & Hillier (1968), Embrey (1970), Roth-Brandell
& Adams (1970) and others have shown that labour can be induced by
PGE_2 and $PGF_{2\alpha}$. Usually, the compounds are administered by intravenous infusion but they are active when administered orally and can also

be applied vaginally and, for second trimester induction of abortion, intra-amniotically. Several other investigators have now confirmed the uterus-activating properties of these two prostaglandins, and from the rapidly growing literature certain features have emerged which I shall comment upon.

While the human uterus shows an increasing sensitivity to oxytocin with advancing gestation, with a dramatic rise at term and during labour, the sensitivity patterns seem to be different for the prostaglandins E_2 and $F_{2\alpha}$. The increase in sensitivity is much less marked, no more than tenfold, a finding which can be utilised therapeutically. The pattern of labour induced with the two compounds at term does not seem to differ significantly from spontaneous labour or labour induced with oxytocin, although a greater tendency to hypertonicity has been described by some investigators.

The sensitivity of the myometrium to prostaglandins is influenced by oestrogens and progestagens. Also the ability to synthesise and release prostaglandins from decidual cells seems to be dependent on the steroid levels. Studies on the relationships between steroids and prostaglandins are clearly needed.

Gillespie *et al.* (1972) have measured plasma oxytocin in 22 women receiving PGE_2 or $F_{2\alpha}$ intravenously for induction of labour. Oxytocin was detected in the plasma of 19 of the 22 women; it was present in 43 per cent of 139 plasma samples.

Karim & Devlin (1967) analysed the content of prostaglandins in human liquor amnii during pregnancy and labour. PGE_1 was found during both pregnancy and labour. PGE_2 was found from the 35th week and during labour, while $PGF_{1\alpha}$ and $F_{2\alpha}$ were present only during active labour. According to Karim (1968) $PGF_{2\alpha}$ is also present in the maternal blood during labour and the concentrations increase with the progression of labour. The concentrations are large enough to support the assumption of a physiological role though Samuelsson (1972), at a recent meeting, stated that according to his calculations the concentrations were 100- to 1000-fold higher than those released from the tissues into the circulation during labour.

Since there is no dramatic increase in the sensitivity of the myometrium to $PGF_{2\alpha}$ at term, it is necessary to postulate the existence of a special mechanism for endogenous production and release operating only at term, in order to explain a physiological role of $PGF_{2\alpha}$ in the initiation and maintenance of uterine contractions. In the absence of major endocrine changes at term, it is difficult to visualise such a mechanism, but there is one observation worth mentioning in this connection. Klöck & Jung (1972) recently studied the effect of stretch on human myometrial strips *in vitro*. They found that stretch increased the amount of PGE in the bath up to 16-fold. When they suspended two strips of tissue from the same uterus, one stretched and the other not, the first exhibited a much higher spontaneous activity than the other; at the end of the experiment the fluid in the first bath contained 33 times as much PGE as the fluid from the second. That a myometrial strip works better at 'optimal length' is well known;

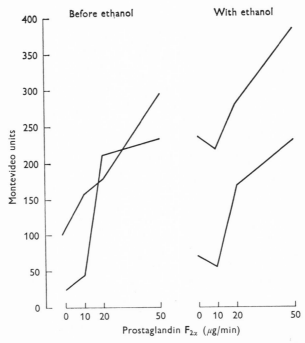

Fig. 4. Dose–response curves for uterine contractions induced by prostaglandin $F_{2\alpha}$ in pregnant baboons within 2 weeks of term, before and immediately after intravenous infusion of ethanol (0·6 g/kg body weight/h, for 2 h). From Lauersen, Raghavan, Wilson, Fuchs & Niemann (1973).

that the activity is associated with a release of prostaglandin is new. If indeed prostaglandins are produced in the myometrial cells during contractions, one could visualise a sensitising and facilitating effect, and possibly even a stimulating effect which could explain the almost 'automatic' activity of the uterus in fully established labour.

Karim (1972) puts forward the following arguments to suggest a role for prostaglandins in labour: (1) $PGF_{1\alpha}$ and $F_{2\alpha}$ are only present in the liquor amnii during labour, and the concentration of PGE_2 is higher during labour than in patients at term but not in labour; (2) PGE_2 and

$F_{2\alpha}$ are present in the maternal circulation only during labour; (3) all four prostaglandins stimulate the pregnant uterus and have been used for induction of labour; (4) intravenous infusion of ethanol inhibits prostaglandin-induced labour. As to the last argument, we have found no effect of ethanol on $PGF_{2\alpha}$-induced labour in the baboon (Fig. 4) and E. M. Coutinho (personal communication, 1972) found no effect of ethanol on

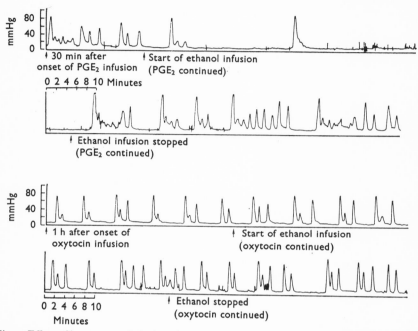

Fig. 5. Effect of intravenous infusion of 10 per cent ethanol on prostaglandin- and oxytocin-induced activity in two patients near term. Reprinted with permission from Karim (1972).

$PGF_{2\alpha}$-induced contractions in pregnant and non-pregnant women. Figure 5 (Karim, 1972) shows an almost immediate effect both at the start of the ethanol infusion and at the end, features we have never seen when ethanol infusions have been given in threatened premature labour.

Obviously, the prostaglandins add a new dimension to the complicated picture of labour, but so far they have not helped to clarify the issue. Many questions, too numerous to be enumerated, need to be asked and answered first.

PROGESTATIONAL HORMONES

The steroid hormones produced in the placenta or the foeto-placental unit, the progestagens and the oestrogens, are identical with those produced by the ovary during the menstrual cycle and in early pregnancy. During the menstrual cycle the uterus is never at rest, but the pattern of contractions is modulated by the levels of ovarian steroids. The strongest contractions are present during menstruation when steroid secretion is declining. In contrast, the labour contractions occur at a time when the levels of progestagens and oestrogens are reaching a maximum. A simple withdrawal effect can therefore be excluded as a possible mechanism for the initiation of labour.

The predominant progestational hormones in man are progesterone, 20α- and 20β-dihydroprogesterone, and 17-hydroxyprogesterone. In many species it has been shown conclusively that the maternal blood levels of progesterone decline markedly before the onset of labour. This is the case in the cow, sheep, rabbit and rat, among others. In the foetus there is a decline in the concentration of progesterone at least in the sheep and the goat, as shown by Liggins, Grieves, Kendall & Knox (1972) and Thorburn

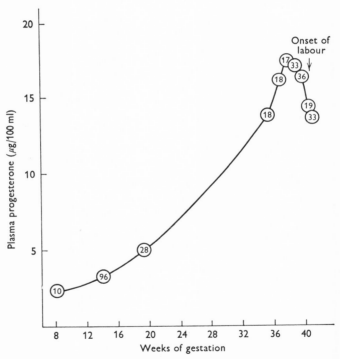

Fig. 6. The progesterone levels in the peripheral plasma of women during pregnancy and spontaneous labour. Numbers in circles refer to the number of replicate analyses. The third trimester and labour values are the results of analysis of 92 plasma samples from 12 hospitalised, obstetrically normal, young, nulliparous women. Reprinted from Csapo et al. (1971).

et al. (1972) respectively. Large doses of progesterone can prevent parturition in several species such as the sheep and the rabbit. Scientific as well as popular beliefs notwithstanding, women do not seem to behave like sheep or rabbits at least in this regard.

A withdrawal of progesterone as a prerequisite for labour in man has been postulated many times, particularly by Csapo and his group, but the evidence is not convincing. The excretion of pregnanediol in the urine,

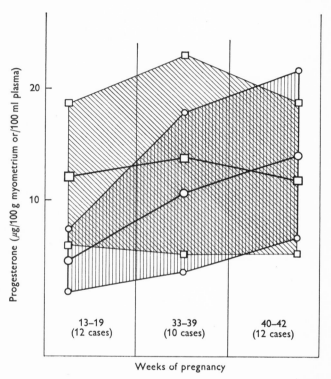

Fig. 7. Mean progesterone values, with the corresponding standard deviations, in myometrium (□), and peripheral venous plasma (○) from the same patients at various stages of pregnancy. Reprinted with permission from Runnebaum & Zander (1971).

the first approach to an evaluation of progestagen metabolism, increases throughout pregnancy, and although it levels off in the last trimester (Shearman, 1959) a fall before the onset of labour is not characteristic. When assays of progesterone in blood became available, it was found that the levels increase throughout pregnancy, and do not fall at the onset of labour (Aitken, Preedy, Eton & Short, 1958; Yannone, McGurdy & Goldfien, 1968; Llauro, Runnebaum & Zander, 1968). An exception is the work of Csapo, Knobil, van der Molen & Wiest (1971) which reported a significant, though not very impressive drop in the concentration of progesterone at the onset of, and during, labour (Fig. 6). The fall does not bring the level below the normal range at this stage of pregnancy. One would expect a more dramatic fall if the progesterone concentration in the blood were a major deciding factor in the mechanism of initiating labour. However, the blood concentrations may not reflect what is going on in the myometrium itself. According to Runnebaum & Zander (1971) the concentration of progesterone in the myometrium remains almost constant throughout human pregnancy (Fig. 7).

Even determination of the concentrations of progesterone in the myometrium may not clarify the situation. The amount of progesterone-binding proteins and the degree of saturation of binding capacity could change without being reflected in the tissue concentration of progesterone, although such changes could have important functional consequences. The work on uterine binding proteins for steroids which is now being conducted in many laboratories is therefore very important.

Roberts & Share (1969, 1970) have found that progesterone inhibits the secretion of oxytocin in response to vaginal stimulation in sheep and goats. Unpublished studies by Fuchs & Lisk (1966) have shown that implantation of progesterone in the median eminence severely affects parturition in rabbits, apparently by the release of oxytocin. Fuchs & Wichmann (1971) found a higher uptake of labelled progesterone in the hypophysis than in other brain areas of pregnant rabbits. The uptake increased with advancing gestation, reaching a peak on day 30 and falling sharply *post partum*. Such studies indicate that progesterone could influence posterior pituitary function at the end of pregnancy.

A very recent report on the progesterone concentration gradient in the human foetal membranes by Pulkkinen & Enkola (1972) is worth mentioning. Foetal membranes were collected from 20 patients at Caesarean section before the onset of labour. The average progesterone concentration was found to be 68.9 ± 4.9 μg/100 g tissue, which is probably of the same order as the concentration in the uterine venous blood at term and much higher than that in the peripheral maternal blood. A comparison of membranes near the placenta and far from the placenta in 12 patients showed values of 85.7 ± 11.7 versus 60.8 ± 6.4 μg/100 g, a difference which is statistically significant ($P < 0.005$). These findings support the concept of a local effect of placental progesterone. It is interesting that the progesterone concentration in the foetal membranes is high, while the concentration in the amniotic fluid is much lower than that in the peripheral maternal blood and, in contrast to the maternal blood levels, declines slowly as pregnancy progresses (Johansson & Jonasson, 1971) (Fig. 8).

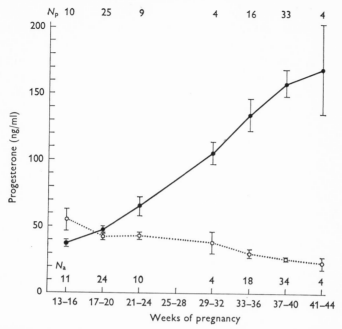

Fig. 8. Progesterone concentrations in plasma (——) and amniotic fluid (........) from the same patients. The values are grouped in 4-week periods. N_p indicates the number of plasma samples in each period; N_a the number of amniotic fluid samples. The vertical bars indicate the s.e.m. Reprinted with permission from Johansson & Jonasson (1971).

The evidence for and against progesterone withdrawal in man has been debated so often that I shall refrain from further comment. My own view is that a withdrawal of progesterone at the onset of human labour has not been clearly demonstrated. One cannot exclude the possibility, however, that the effect of progesterone is somehow prevented at the onset of labour, by displacement, inhibition, inactivation, or other mechanisms.

THE FUNCTIONAL ROLE OF OESTROGENS IN LABOUR

While the oestrogens, notably oestriol, provide a good indication of foetal condition, the role of oestrogens in the mechanism of labour is poorly understood. Studies *in vitro* indicate that the oestrogen-dominated myometrium shows little spontaneous activity but a strong reactivity to oxytocin and other oxytocic agents. But in pregnancy, where the oestrogen levels are rising steeply, the 'spontaneous' activity increases with advancing gestation.

Pinto, Fisch, Schwarcz & Montuori (1964) considered oestradiol to be oxytocic *in vitro* and they, as well as Järvinen, Luukkainen & Väistö (1965), claimed to have increased the sensitivity of the pregnant uterus to induction of labour with oxytocin. Klopper & Dennis (1962) on the other hand, failed to demonstrate any difference in the length of the induction-to-delivery interval in three groups of subjects treated with stilboestrol, oestriol, and placebo, respectively. Later, Klopper, Dennis & Farr (1969) found a stimulatory effect of oestriol injected into the amniotic cavity.

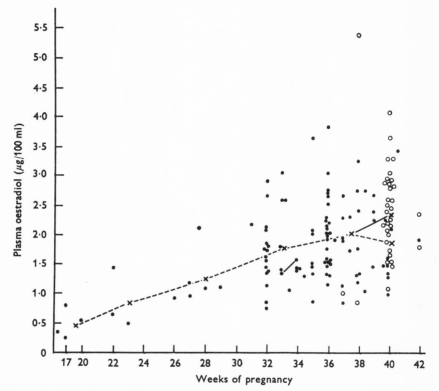

Fig. 9. Individual oestradiol concentrations in maternal plasma in uncomplicated pregnancies. The broken line joins the mean values for different periods of gestation of patients not in labour. The solid line shows the mean value for term patients in early labour. The open circles are individual values from patients in labour. Reprinted with permission from Sybulski & Maughan (1972).

With the development of competitive protein-binding assays and radio-immunoassays for oestrogens, more information should become available about the plasma levels of oestrogens during pregnancy. Quite recently, Sybulski & Maughan (1972) published a study of the levels of oestradiol in the maternal blood in the second half of pregnancy (Fig. 9). It is interesting that the levels were higher, on the average, in women in labour than in women at term but not yet in labour. However, while the difference is significant ($P<0\cdot05$), the range for each group is wide with considerable overlapping. Oestriol levels at term, determined by a number of investigators, are of the same order of magnitude.

YoungLai, Effer & Pelletier (1971) have measured the concentration of oestrogens in the amniotic fluid during gestation. Their results are shown in Fig. 10, which illustrates that the levels of 'immunoreactive oestrogens',

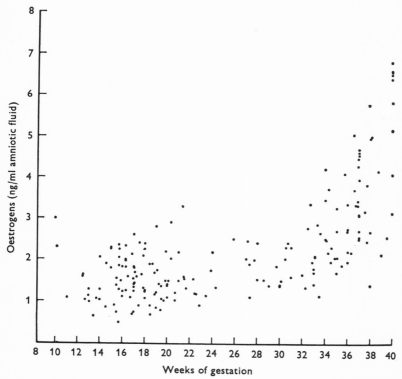

Fig. 10. Concentrations of immunoreactive oestrogens (mainly oestrone and oestradiol-17α and -17β) in the amniotic fluid of women at different stages of pregnancy. Reprinted with permission from YoungLai *et al.* (1971).

mainly oestrone and oestradiol-17α and -17β, are about one fifth of the oestradiol levels in the maternal plasma.

If the placental steroids determine the sensitivity of the myometrium to oxytocic agents, and if oestrogens and progestagens have opposite effects, a study of the ratio of oestrogens to progestagens may be rewarding. According to Klopper, determination of the ratio as expressed by the oestriol and pregnanediol excretions in the urine did not provide any clue, perhaps not surprisingly, considering how poorly the urine values reflect the plasma concentrations.

The oestrogens are the most potent of the steroids and they affect nearly all systems and tissues. Their concentrations increase greatly during pregnancy and it is inconceivable that this is without significance for the onset of labour.

CORTICOSTEROIDS

Thanks to the work of Liggins and his coworkers (1967, 1968, 1969, 1971, 1972) on the role of foetal corticosteroids, work which is now being pursued in many parts of the world, there can be no doubt that the corticosteroids have a significant function in the process of labour. This work has taught us more about parturition in the sheep than in almost any other species, but since Chez, Hutchinson, Salazar & Mintz (1970) studied the effects of foetal hypophysectomy in the rhesus monkey with similar results, it is clear that foetal corticosteroids must play an important role in primates also.

Giroud and his group have studied the concentrations of corticosteroids and their sulphates in maternal and foetal blood. Figure 11, modified from

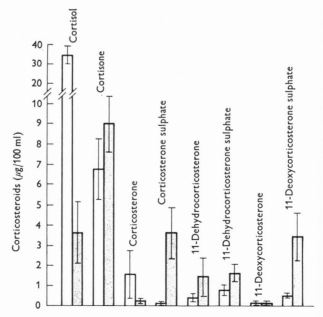

Fig. 11. Concentrations of corticosteroids in the maternal blood of women at term delivery (open columns) and cord blood (shaded columns) from seven normal women. The vertical lines indicate the s.e.m. Based upon data from Giroud (1971).

Giroud (1971), shows concentrations of corticosteroids at delivery. How the levels changed in the days before the onset of labour, particularly in the foetus, is not known. In a communication to the recent International Congress of Endocrinology (Washington, 1972), Dr Beverley Murphy of Montreal reported that the levels of corticosteroids in cord plasma were higher in women in whom labour was spontaneous than in women in whom labour had been induced. This would seem to indicate that the concentrations go up before the onset of labour. Does that mean the spontaneous labour is triggered by a rise in the foetal levels of cortico-steroids? The observation by Murphy also seems to hold for aldosterone.

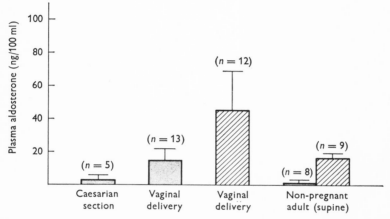

Fig. 12. Concentrations of aldosterone in maternal plasma of women at the time of delivery by elective Caesarean section and by the vaginal route. The columns represent the mean plasma aldosterone concentrations in ng/100 ml and the vertical lines the s.d. Stippled columns = normal sodium; hatched columns = low sodium ± diuretics. Re-printed with permission from Beitins et al. (1972).

Figure 12 from Beitins et al. (1972) shows that the maternal levels at elective Caesarean section were considerably lower than at vaginal delivery. The same was the case for the concentrations in cord blood which, on average, were higher than the maternal levels. The concentrations in both maternal and cord blood were higher in patients on reduced sodium intake than in normal patients.

My time is up, and with this and other questions left to be answered in the course of the symposium, I shall conclude my review. You will agree that while I was too long-winded, I omitted more than I included. A number of hormones were not even mentioned and many hypotheses were left to be brought out later. If my presentation had been comprehensive, I would have failed. If I have been provocative, I have at least accomplished something.

REFERENCES

AHLQUIST, R. P. (1948). The study of the adrenotropic receptors. *Am. J. Physiol.* **153**, 586–600.

AITKEN, E. H., PREEDY, J. R. K., ETON, B. & SHORT, R. V. (1958). Oestrogen and progesterone levels in foetal and maternal plasma at parturition. *Lancet*, ii, 1096–1099.

BASHORE, R. A. (1972). Studies concerning the radioimmunoassay for oxytocin. *Am. J. Obstet. Gynec.* **113**, 488–496.

BEITINS, I. Z., BAYARD, F., LEVITSKY, L., ANCES, I. G., KOWARSKI, A. & MIGEON, C. J. (1972). Plasma aldosterone concentration at delivery and during the newborn period. *J. clin. Invest.* **51**, 386–394.

BRANDA, L. A. & FERRIER, B. M. (1971). Degradation of oxytocin by human placental tissue. *Am. J. Obstet. Gynec.* **109**, 943–947.

CHARD, T. (1972). The posterior pituitary in human and animal parturition. *J. Reprod. Fert.* Suppl. **16**, 121–138.

CHARD, T., BOYD, N. R. H., EDWARDS, C. R. W. & HUDSON, C. N. (1971). Release of oxytocin and vasopressin by the human foetus during labour. *Nature, Lond.* **234**, 352–354.

CHARD, T., BOYD, N. R. H., FORSLING, M. L., MCNEILLY, A. S. & LANDON, J. (1970). The development of a radioimmunoassay for oxytocin. The extraction of oxytocin from plasma, and its measurement during parturition in human and goat blood. *J. Endocr.* **48**, 223–234.

CHEZ, R. A., HUTCHINSON, D. L., SALAZAR, H. & MINTZ, D. H. (1970). Some effects of fetal and maternal hypophysectomy in pregnancy. *Am. J. Obstet. Gynec.* **108**, 643–650.

COCH, J. A., BROVETTO, J., CABOT, H. M., FIELITZ, C. A. & CALDEYRO-BARCIA, R. (1965). Oxytocin equivalent activity in the plasma of women in labour and puerperium. *Am. J. Obstet. Gynec.* **91**, 10–17.

CSAPO, A. I., KNOBIL, E., VAN DER MOLEN, H. J. & WIEST, W. G. (1971). Peripheral plasma progesterone levels during human pregnancy and labour. *Am. J. Obstet. Gynec.* **110**, 630–632.

EMBREY, M. P. (1970). Induction of labour with prostaglandins E_1 and E_2. *Br. med. J.* ii, 256–258.

FALCK, B. (1962). Observations on the possibilities of the cellular localization of monoamines by a fluorescence method. *Acta physiol. scand.* **56** (Suppl. 197), 1–25.

FITZPATRICK, R. J. & WALMSLEY, C. F. (1965). The release of oxytocin during parturition. In *Advances in Oxytocin Research*, ed. J. H. M. Pinkerton, pp. 57–71. London: Pergamon Press.

FUCHS, A.-R. & COUTINHO, E. M. (1971). Suppression of uterine activity during menstruation by expansion of the plasma volume. *Acta endocr., Copenh.* **66**, 183–192.

FUCHS, A.-R., COUTINHO, E. M., XAVIER, R., BATES, P. E. & FUCHS, F. (1968). Effect of ethanol on the activity of the non-pregnant human uterus and its reactivity to neurohypophyseal hormones. *Am. J. Obstet. Gynec.* **101**, 997–1000.

FUCHS, A.-R. & LISK, R. D. (1966). Quoted by Fuchs, A.-R. (1966) in Studies on the control of oxytocin release at parturition in rabbits and rats. *J. Reprod. Fert.* **12**, 418.

Fuchs, A.-R. & Wichmann, K. (1971). The uptake of ³H-progesterone and its metabolites by various brain areas of pregnant rabbits. *Acta endocr., Copenh.* **67** (Suppl. 155), 22.

Fuchs, F., Fuchs, A.-R., Poblete, V. F., Jr & Risk, A. (1967). Effect of alcohol on threatened premature labour. *Am. J. Obstet. Gynec.* **99**, 627–637.

Gillespie, A., Brummer, H. C. & Chard, T. (1972). Oxytocin release by infused prostaglandin. *Br. med. J.* i, 543–544.

Giroud, C. J. P. (1971). Aspects of corticosteroid biogenesis and metabolism during the perinatal period. *Clin. Obstet Gynec.* **14**, 745–762.

Holm, L. W. (1967). Prolonged pregnancy. *Advances in veterinary Science,* **11**, 159–205.

Hoppenstein, J. M., Miltenberger, F. W. & Moran, W., Jr (1968). The increase in blood levels of vasopressin in infants during birth and surgical procedures. *Surg. Gynec. Obstet.* **127**, 966–974.

Järvinen, P. A., Luukkainen, T. & Väistö, L. (1965). The effect of oestrogen treatment on myometrial activity in late pregnancy. *Acta obstet. gynec. scand.* **44**, 258–264.

Johansson, E. D. B. & Jonasson, L.-E. (1971). Progesterone levels in amniotic fluid and plasma from women. *Acta obstet. gynec scand.* **50**, 339–343.

Karim, S. M. M. (1968). Appearance of prostaglandin $F_{2\alpha}$ in human blood during labour. *Br. med. J.* iv, 618–621.

Karim, S. M. M. (1972). Physiological role of prostaglandins in the control of parturition and menstruation. *J. Reprod. Fert.* Suppl. **16**, 105–119.

Karim, S. M. M. & Devlin, J. (1967). Prostaglandin content of amniotic fluid during pregnancy and labour. *J. Obstet. Gynaec. Br. Commonw.* **74**, 230–234.

Karim, S. M. M., Trussell, R. R., Patel, R. C. & Hillier, K. (1968). Response of the pregnant human uterus to prostaglandin $F_{2\alpha}$ induction of labour. *Br. med. J.* iv, 621–628.

Klöck, F. K. & Jung, H. (1972). Extraktion und dünnschichtchromatographische Untersuchungen des 'Uterus-Hemmstoff'. In *Methoden der Pharmakologischen Geburtserleichterung und Uterus-Relaxation,* ed. H. Jung, pp. 177–183. Stuttgart: G. Thieme.

Klopper, A. I. & Dennis, K. J. (1962). Effect of oestrogens on myometrial contractions. *Br. med. J.* ii, 1157–1159.

Klopper, A. I., Dennis, K. J. & Farr, V. (1969). Effect of intra-amniotic oestriol sulphate on uterine contractions. *Br. med. J.* ii, 786–789.

Kumaresan, P., Anandarangam, B. & Vasicka, A. (1972). Studies of human oxytocin by radioimmunoassay. In *Abstracts of the IVth International Congress of Endocrinology, Washington D.C., June 1972,* Abstr. no. 256, p. 206. Amsterdam: Excerpta Medica Foundation.

Lauersen, N. H., Raghavan, K. S., Wilson, K. H., Fuchs, F. & Niemann, W. H. (1973). Effect of prostaglandin $F_{2\alpha}$, oxytocin, and ethanol on the uterus of the pregnant baboon. *Am. J. Obstet. Gynec.* **115**, 912–918.

Liggins, G. C. (1968). Premature parturition after infusion of corticotrophin or cortisol into foetal lambs. *J. Endocr.* **42**, 323–329.

Liggins, G. C. (1969). Premature delivery of foetal lambs infused with glucocorticoids. *J. Endocr.* **45**, 515–523.

Liggins, G. C. & Grieves, S. A. (1971). Possible role for prostaglandin $F_{2\alpha}$ in parturition in sheep. *Nature, Lond.* **232**, 629–631.

Liggins, G. C., Grieves, S. A., Kendall, J. C. & Knox, B. S. (1972). The physiological roles of progesterone, oestradiol-17β and prostaglandin $F_{2\alpha}$ in the control of ovine parturition. *J. Reprod. Fert.* Suppl. **16**, 85–103.

Liggins, G. C., Kennedy, P. C. & Holm, L. W. (1967). Failure of initiation of parturition after electrocoagulation of the pituitary of the fetal lamb. *Am. J. Obstet. Gynec.* **98**, 1080–1086.

LLAURO, J. L., RUNNEBAUM, B. & ZANDER, J. (1968). Progesterone in human peripheral blood before, during, and after labor. *Am. J. Obstet. Gynec.* **101**, 867–873.

NAKANISHI, H., McLEAN, J., WOOD, C. & BURNSTOCK, G. (1969). The role of sympathetic nerves in control of the non-pregnant and pregnant human uterus. *Lying-in, J. reprod. Med.* **2**, 20–33.

OWMAN, C., ROSENGREN, E. & SJÖBERG, N.-O. (1967). Adrenergic innervation of the human female reproductive organs: a histochemical and chemical investigation. *Obstet. Gynec., N.Y.* **30**, 763–773.

PINTO, R. M., FISCH, L., SCHWARCZ, R. & MONTUORI, E. (1964). Action of estradiol-17β upon uterine contractility and the milk-ejecting effect in the pregnant woman. *Am. J. Obstet. Gynec.* **90**, 99–114.

PULKKINEN, M. O. & ENKOLA, K. (1972). The progesterone gradient of the human fetal membranes. *Int. J. Gynaec. Obstet.* **10**, 93–94.

ROBERTS, J. S. & SHARE, L. (1969). Effects of progesterone and estrogen on blood levels of oxytocin during vaginal distension. *Endocrinology* **84**, 1076–1081.

ROBERTS, J. S. & SHARE, L. (1970). Inhibition by progesterone of oxytocin secretion during vaginal stimulation. *Endocrinology* **87**, 812–815.

ROTH-BRANDELL, U. & ADAMS, M. (1970). An evaluation of the possible use of prostaglandin E_1, E_2 and $F_{2\alpha}$ for induction of labour. *Acta obstet. gynec. scand.* **49** (Suppl. 5), 9–17.

RUNNEBAUM, B. & ZANDER, J. (1971). Progesterone and 20α-dihydroprogesterone in human myometrium during pregnancy. *Acta endocr., Copenh.* Suppl. **150**, 5–50.

SAMUELSSON, B. (1972). Brook Lodge Symposium on Prostaglandins, June 12–14 (discussion). Upjohn Company.

SHEARMAN, R. P. (1959). Some aspects of the urinary excretion of pregnanediol in pregnancy. *J. Obstet. Gynaec. Br. Commonw.* **64**, 1–11.

SJÖBERG, N.-O. (1967). The adrenergic transmitter of the female reproductive tract: distribution and functional changes. *Acta physiol. scand.* Suppl. **305**, 5–32.

SYBULSKI, S. & MAUGHAN, G. B. (1972). Maternal plasma estradiol levels in normal and complicated pregnancies. *Am. J. Obstet. Gynec.* **113**, 310–315.

THORBURN, G. D., NICOL, D. H., BASSETT, J. M., SHUTT, D. S. & COX, R. I. (1972). Parturition in the goat and sheep: changes in corticosteroids, progesterone. *J. Reprod. Fert.* Suppl. **16**, 61–84.

VORHERR, H. (1972). ADH and oxytocin in blood and urine of gravidas and parturients. Abstr. no. 38. Society for Gynecological Investigation: Annual Meeting, March.

WANSBROUGH, H., NAKANISHI, H. & WOOD, C. (1968). The effect of adrenergic receptor blocking drugs on the human uterus. *J. Obstet. Gynaec. Br. Commonw.* **75**, 189–198.

YANNONE, M. E., McGURDY, J. R. & GOLDFIEN, A. (1968). Plasma progesterone levels in normal pregnancy, labor, and the puerperium. *Am. J. Obstet. Gynec.* **101**, 1058–1061.

YOUNGLAI, E. V., EFFER, S. B. & PELLETIER, C. (1971). Amniotic fluid progestins and estrogens in relation to length of gestation. *Am. J. Obstet. Gynec.* **111**, 833–839.

ZUSPAN, F. P. (1970). Urinary excretion of epinephrine and norepinephrine during pregnancy. *J. clin. Endocr. Metab.* **30**, 357–360.

FOETAL PHYSIOLOGY AND THE ONSET OF LABOUR

By G. S. DAWES

INTRODUCTION

In the past few years it has become clear that in the sheep and goat the foetus is responsible for the normal initiation of labour through a mechanism which involves the hypothalamus and the pituitary, with the release of glucocorticoids from the foetal adrenal cortex and a terminal decrease in circulating progesterone levels and rapid rise in plasma oestrogens and prostaglandin $F_{2\alpha}$ (Liggins, 1969; Liggins, Grieves, Kendall & Knox, 1972; Thorburn et al. 1972). Such evidence as is available for man suggests that a similar mechanism may operate. The evidence available is insufficient to judge of other species, though there is a reasonable presumption that the same principle may apply.

Several questions then arise. First, what is the central mechanism which operates to start the process in the hypothalamus with a timing that, in the sheep, is accurate to ± 2 days in 147? Is there a central time clock, or does the initiating stimulus depend upon the maturation of some peripheral organ system? Secondly, we must consider whether there are factors which explain the propensity of mothers of several species to give birth during the hours of darkness. Thirdly, we may enquire whether there are physiological events in the foetus associated with the hormonal changes preceding labour. To try to answer some of these questions one has to look more widely at the general physiology of the foetus *in utero*. Fortunately this has become more accessible to investigation through the development of continuous recording methods with chronic implantation of catheters, electrodes and electromagnetic flowmeters.

FOETAL BREATHING MOVEMENTS *IN UTERO*

Foetal breathing movements *in utero* occur normally under favourable physiological conditions, i.e. in the absence of acidaemia or hypoxaemia, in lambs and rabbits (Dawes et al. 1970, 1972; Merlet, Hoerter, Devilleneuve & Tchobroutsky, 1970) and in human foetuses *in utero* (Boddy & Robinson, 1971). In sheep (delivered in acute experiments into a saline

[25]

bath) these breathing movements are present from an early gestational age (40 days); as early as that at which skeletal muscular movements can first be observed. The breathing movements are episodic in character and discontinuously present for up to 40 per cent of the time. Most of these breathing movements consist of 'rapid irregular breathing', i.e. movements which vary irregularly in rate and depth. The average rate and depth increase with gestational age from <1 Hz and <1 mmHg intratracheal pressure at 40 days of gestation, to 2–4 Hz and sometimes more than 30 mmHg near term. These rapid irregular breathing movements are associated with rapid-eye-movement sleep. The evidence for this depends first on observations made over periods of many hours on foetuses delivered by Caesarean section under epidural anaesthesia directly into a warm saline bath with intact umbilical cords. The gestational age of the foetuses ranged upwards from 0·78 of term (115 days gestation). Secondly, eye movements were recorded *in utero* from electrodes implanted along the superior orbital ridge, and thirdly, records were made of the foetal electrocorticogram from electrodes implanted through the parietal bone on to the dura, and encased in dental acrylic. These observations have shown that the foetal lamb spends more than a third of the 24 h in rapid-eye-movement sleep (see also Ruckebusch, 1971, 1972). Rapid irregular breathing movements only occur during this time.

There are also other foetal breathing movements. About 10 per cent of the time relatively deep but less frequent inspiratory efforts (as judged from the intratracheal pressure record) are present at a rate of 1–4/min. These have been described as gasping movements, but it is possible that they may represent sighing. They are not associated with any particular phase of sleep or of cortical activity. Neither rapid irregular breathing movements nor gasping are normally associated with variations from the normal in the foetal blood gas values. However if the physiological condition of a foetus *in utero* deteriorates rapidly for any cause (e.g. from haemorrhage through displacement of the catheter from the carotid artery, or from asphyxia through the umbilical cord being twisted around flowmeter leads) death is preceded by a series of deep asphyxial gasping movements which succeed each other at regular intervals.

There are other types of foetal thoracic muscular movement, as inferred from measurements of tracheal pressure and flow. For instance from time to time there may be single, or a succession of a few rapidly repeated expiratory efforts suggestive of coughing or bleating. And there are also occasional changes in intrathoracic pressure without any equivalent change in amniotic fluid pressure.

In summary we have a picture of a wide range of respiratory muscular movements, similar to those present after birth but differing in certain features. The most common of these and in many ways the most interesting are those described as rapid irregular breathing movements associated with rapid-eye-movement sleep. It is especially interesting that these rapid irregular breathing movements are present from such an early stage of gestation for the cerebral cortex is still anatomically and functionally poorly developed; continuous electrocortical activity is not present before about 75 days of gestational age (Bernhard & Meyerson, 1968). It is also pertinent that newborn human infants spend much of their time in rapid-eye-movement sleep; this is associated with an irregularity of respiratory movement which may be due to the same mechanism as that responsible for rapid irregular breathing in lambs *in utero*.

Preliminary observations on human infants *in utero*, using an ultrasonic method for detecting changes in thoracic shape, suggest that rapid irregular breathing movements are present more often than in lambs (up to 70 per cent of the time) but at a lower frequency, about 70/min (Boddy & Mantell, 1972).

The rate and depth of rapid irregular breathing movements has been measured from foetal tracheal pressure and flow records. These show that a very moderate degree of hypoxaemia, a fall of 5–7 mmHg P_{O_2} in carotid arterial blood unaccompanied by acidaemia abolishes rapid irregular breathing movements. This degree of hypoxaemia does not cause any large change in electrocortical activity; if anything there is more large-voltage slow activity. Hypercapnia (induced by giving the ewe a gas mixture containing 4–6 per cent carbon dioxide to breathe) causes an increase in the height and depth of foetal breathing measurements. There is a definite increase in the minute volume of breathing (as computed from integrated tracheal flow records) for an approximate 10 mmHg rise in foetal carotid P_{CO_2}. This indicates a much lower sensitivity of the central chemoreceptor mechanism of the foetus *in utero* than after birth. Other pathological changes such as hypoglycaemia or infection diminish or arrest foetal breathing. The interaction of these variables has yet to be explored.

Effects upon the circulation

With foetal breathing movements *in utero* there are changes in the circulation. Deep inspiratory efforts are mechanically transferred to the heart and great vessels, because the chest is entirely filled with fluid. Consequently rapid irregular foetal breathing is associated with a gross irregularity of the pulse, whether this is recorded as pressure in the carotid artery or as flow in the descending aorta (measured by a cuff electromagnetic flowmeter).

There is also the possibility that, as in the adult, the central regulation of the circulation may change with different phases of sleep; this is a problem which remains to be explored.

In summary, while arterial pressure and heart rate are relatively invariant when the foetus is not making breathing movements, at other times there is an episodic variation in association with foetal breathing, present throughout the last half of gestation. The magnitude of this variation increases with gestational age. At term it can be large, with rises of arterial pressure of the order of 20 mmHg ($\sim$ 30 per cent).

It is pertinent to mention at this point the evidence from numerous sources of the relative competence of the circulation at birth in all the species so far studied. This has been related in the past to the progressive development of the autonomic nervous system towards the end of gestation in sheep, together with the known functioning of the baroreceptor and aortic chemoreceptor reflexes. But there is now evidence for the relatively greater importance of the renin–angiotensin system in the period immediately after birth (Broughton Pipkin, Mott & Roberton, 1971; Mott, 1973). In foetal lambs near term this system can be activated by injection of frusemide (Trimper & Lumbers, 1972) or, in acute experiments, by haemorrhage (J. C. Mott, F. Broughton Pipkin & E. Lumbers, unpublished). It remains to be seen to what extent this system plays a role in the normal control of foetal arterial pressure *in utero*, and whether aldosterone secretion may be activated.

Tracheal fluid flow

The normal physiology of tracheal fluid flow from the alveoli through the larynx is of interest because this must be the principal means by which pulmonary surfactant reaches the amniotic fluid. It is comparatively easy to measure the net inflow or outflow of fluid by integrating the output of an electromagnetic flow meter implanted in the trachea. (The arrangement differs from that described above for measuring gross to-and-fro tracheal fluid flow, in that for the latter the flowmeter record is full wave rectified before integration.) The results show that there is occasionally a small inflow of fluid into the trachea from the pharynx associated with the most vigorous rapid irregular breathing movements; but this inflow is rarely more than a few ml. There is a net outward fluid flow which averages 100 ml/kg/day near term. Part of this fluid transfer takes place in very small outward flows of a few ml. The larger part occurs in 6–12 episodes a day, during which there is a large outward tracheal flow (i.e. from the lungs to the pharynx) which can be as much as 30–40 ml over a period of 5 min.

These episodes are not associated with any particular pattern of breathing movements. They are associated with a considerable rise of arterial pressure (although only 10–20 mmHg, this is large for the foetus). They may also be associated with swallowing movements (see below). In four lambs in which net outward tracheal flow has been systematically analysed there is evidence that both the outward flow and the number and size of the episodic gushes falls 2 to 3 days before the onset of labour.

SWALLOWING

During the past year Bradley & Mistretta (1973) have studied the physiology of swallowing in the foetal lamb *in utero* by measuring oesophageal flow from a cannulated electromagnetic flowmeter implanted in the neck. Swallowing occurred in brief irregular episodes, each lasting several minutes, at intervals of some hours. The number of such swallowing episodes varies irregularly day by day, and is normally four to ten. Each episode lasts for 2–5 min and is associated with the movement of large quantities of fluid down the oesophagus: up to a maximum of 200 ml in a lamb near term. Each swallowing episode is accompanied by a rise of arterial pressure of up to about 20 mmHg. The number of episodes of swallowing and the daily volume of fluid swallowed decreased 2 to 3 days before the onset of labour.

Neither swallowing nor net outward tracheal flow show evidence of a circadian rhythm.

CIRCADIAN RHYTHMS IN FOETAL BREATHING AND THE ELECTROCORTICOGRAM

In the past year Boddy, Dawes & Robinson (1973) have shown that there is a circadian rhythm in breathing. A search was made for this phenomenon when breathing movements in the foetus were first observed (Dawes *et al.* 1972). The records then available were not quantitative. With the introduction of a method for measuring the minute volume of foetal breathing *in utero* (as described above) it became obvious that there was a circadian rhythm. The minute volume of tracheal fluid flow increases during the late evening to reach a peak value 2·0–2·8 times that observed during the trough. The peak is relatively sharp lasting not more than 2 h between 19.00 and 22.00 h. Thereafter the minute volume of foetal breathing falls rapidly to a trough which lasts for 3–5 h between 03.00 and 08.00 h. The peak occurs at an earlier time in winter than it does in summer. All the observations so far have been made with the sheep in a metabolism cage inside the laboratory.

There was also a highly significant circadian variation in the numbers of minutes per hour during which rapid irregular breathing movements were observed on the tracheal pressure record. An examination of the biparietal electrocorticogram in seven pregnant sheep from 124–137 days of gestation showed that there was a highly significant circadian variation in the numbers of minutes in each hour during which predominantly low-voltage rapid activity was present. The normal transition between predominantly low-voltage rapid cortical activity (associated with rapid-eye-movement sleep or wakefulness) and large-voltage slow activity is not readily distinguishable before 124 days of gestational age.

The point has already been made that evidence of spontaneous respiratory muscular activity is present in the foetal lamb long before the cerebral cortex is fully developed. The present evidence shows that there is a circadian rhythm in foetal breathing movements, measurable quantitatively from 110 days of gestation onwards. (At earlier gestational ages tracheal fluid flow cannot be used as a measure of breathing movements because the alveoli are still solid and the lung inexpansible.) So there is a period between 110 and 124 days at which we can be sure that there is a circadian rhythm of breathing, but variations in electrocortical activity associated with a different phase of sleep have not been detected by the relatively crude methods at present available. The question now arises as to whether the circadian rhythm of breathing is dependent upon the development of the cerebal cortex. This seems unlikely and we suppose, on present evidence, that the rhythm arises from sub-cortical, perhaps brain-stem, structures.

It has already been noted (Dawes *et al.* 1972) that the onset of labour induced by dexamethasone infusion is not necessarily associated with a gross change in rapid irregular breathing movements. But these observations were made at a time when quantitative measurement of foetal breathing movements had not been developed. From time to time in the past two years it was found that foetal breathing movements were reduced in frequency and magnitude a day or two before the onset of premature or term labour. One of the difficulties in interpreting such observations is that in some lambs in which premature labour has occurred, there has also been a small fall of foetal carotid arterial P_{O_2}, which would of itself have predisposed to the reduction of foetal breathing *in utero*. Yet this was not always so; there was sometimes a reduction in breathing without evidence of hypoxaemia. The onset of labour is neither inevitably preceded by, nor always associated with, foetal hypoxaemia. The possible causal connection between foetal hypoxaemia and the onset of labour will be discussed in more detail below.

So far the results available are not sufficiently numerous to be able to describe with certainty the effect of the onset of labour upon the magnitude of the circadian rhythms of foetal breathing and electrocortical activity. Buddingh, Parker, Ishizaki & Tyler (1971) have described a circadian rhythm in foetal urinary water excretion.

We may now consider the possible mechanisms involved in a circadian rhythm in the foetus. These may depend either on a direct effect of environmental changes (e.g. sound) upon the foetus, or on the presence of an innate foetal rhythm which becomes locked to diurnal variations in the environment of the mother, or it may depend upon the presence of the rhythm in the mother (thermal, metabolic or hormonal) communicated to the foetus either physically or by placental transfer. Of these possibilities we can say at once that any circadian thermal variation within the mother will be transferred passively to the foetus. The foetal body temperature is normally maintained about 0·5 °C above that of the mother and varies *pari passu* with it. It is less likely that the phenomenon results from direct transfer of hormones across the placenta; the evidence for placental transfer of adrenocorticotrophic hormone (ACTH) or the thyroid hormones in physiological concentrations is poor. The observations of Espinar, Hart & Beaven (1972) suggested there were no clear diurnal variations in corticosteroid levels in adult sheep, though they are well authenticated in other species. However, recently, McNatty, Cashmore & Young (1972) have detected diurnal variations in plasma corticosteroids in anoestrous ewes. The most likely possibilities are thermal variations, metabolic changes, the transfer of other hormones of low molecular weight, and also variations in maternal placental blood flow.

A daily variation in the circulatory levels of foetal hormones is also worth consideration. For instance if there were a variation in foetal corticosteroid levels with time over the 24 h, this might explain the variations in electrocortical activity (Gillan, Jacobs, Fram & Snyder 1972). Such variations might not be easy to detect at the normally low levels of such hormones in the foetus before the onset of labour.

THE EFFECT OF HYPOXAEMIA ON LIBERATION OF ADRENOCORTICOTROPHIC HORMONE IN THE FOETAL LAMB

During the course of experiments designed to test the effect of hypoxaemia on foetal breathing *in utero* our suspicions were aroused that hypoxaemia might lead to the induction of labour. The experiments had been conducted in the following ways. The arterial pressure, amniotic fluid pressure and breathing movements were recorded continuously from foetal lambs *in utero* with chronically implanted catheters, and usually also a tracheal electromagnetic flowmeter. At least a 57 h interval was allowed post-operatively. Hypoxaemia was not induced until clear recordings had been obtained for 6 h previously. Hypoxaemia was then induced by giving the mother a low oxygen mixture (9 per cent oxygen in nitrogen) containing enough carbon dioxide to prevent hypocapnia as a result of maternal hyperventilation. The experiments, which were usually successful, were designed to produce a fall in foetal carotid P_{O_2} of 5–7 mmHg in the absence of significant changes in P_{CO_2}. In some lambs there was no significant variation in pH over the 1 h period during which hypoxaemia was maintained. This relatively long period of hypoxaemia was the minimum which could be used to analyse the effect of breathing movements because of their episodic character. The experiments entailed the removal of an appreciable fraction of the total foetal blood volume for analysis of blood gases and other blood constituents, yet the quantity removed rarely amounted to 10 ml in any one day (about 4 per cent of the blood volume in a foetus weighing 2 kg at 125 days of gestation). It was thought inadvisable to subject the ewe to a further episode of hypoxaemia on the following day; therefore at least a day's interval was allowed between each experiment. Whatever the condition of the ewe and foetus initially the ewe went into labour after only two or three episodes of hypoxaemia, within 7–10 days. While we were considering these facts Alexander, Britton, Forsling, Nixon & Ratcliffe (1972) at the 1971 Symposium of the Society for Endocrinology, Cardiff, reported acute observations on foetal lambs (of 90–140 days of gestation) subjected to hypoxia by giving the mother 10 per cent oxygen in nitrogen to breathe (Alexander *et al.* 1972). This reduced the umbilical arterial P_{O_2} of the foetus to about 10 mmHg; whereas the ewe maintained a constant blood pH or became alkalaemic from overbreathing, the foetus became acidaemic. There was a large rise in foetal plasma ACTH concentration over the period of hypoxia (80–120 min). The initial values for foetal plasma ACTH were variable, perhaps as a result of the acute operative procedure; the period of observation did not allow for the return to normal levels after the cessation of hypoxia. With these reservations there seemed to be a *prima facie* case for believing that hypoxia might indeed induce liberation of ACTH from the foetal pituitary.

To test the validity of this hypothesis observations have been made on 12 pregnant ewes, including three used for control observations (K. Boddy, C. Jones, C. Mantell, J. Ratcliffe & J. Robinson, unpublished). In the nine exposed to hypoxaemia (the ewe receiving 9 per cent oxygen plus 3 per cent carbon dioxide in nitrogen) there was a fall in foetal carotid P_{O_2} of 3–11 mmHg accompanied by a two- to eightfold increase in plasma ACTH concentration. There was sometimes a small transient rise in the maternal ACTH concentration but never of the order seen in the foetus. The foetal plasma ACTH level fell rapidly within 10 min of the cessation of hypoxaemia and was usually close to its initial level 1 h later. In two lambs there was a very clear rise in foetal arterial ACTH in spite of the fact that hypoxaemia was not accompanied by a significant change in foetal arterial pH values. Although several of the sheep were in the early stages of labour, hypoxaemia still caused the same response: a large increase in foetal plasma ACTH, even in instances where the level was already raised. Preliminary results suggest that small haemorrhages, of the size required for blood sampling during these experiments, were not associated with a rise in foetal plasma ACTH levels in the absence of hypoxaemia. While it is evident that we need to know more about the normal variations of plasma ACTH levels in the foetus, and of their changes in the experimental situations to which the ewe is exposed, these observations support the theory that there is a causal connection between hypoxaemia and liberation of ACTH into the foetal circulation.

This evidence is supported by a more general consideration. Experiments involving the implantation of catheters, electrodes and flowmeters chronically into foetal sheep often ultimately result in premature delivery. The chance of this is increased if there are twins and if several electrodes or transducers are implanted, necessitating a longer operation. However the length of the operative procedure is unlikely to be the sole determinant of the onset of premature labour. It seems improbable that it would bring about the onset of labour more than, say, 7 days post-operatively. In many foetal lamb preparations, observed over periods of time considerably longer than 7 days, an association has been noticed between a fall in carotid arterial P_{O_2} and the onset of the labour, whether or not the ewe has been exposed to hypoxaemia. Yet hypoxaemia is by no means the only possible factor which must be considered.

CONCLUSION

The observations described demonstrate that the foetal lamb *in utero* is much more active physiologically than has previously been supposed. Its behaviour changes swiftly from quiet sleep to rapid-eye-movement sleep, and there are occasional periods which may amount to something like wakefulness. Underlying this changing pattern of activity is a circadian rhythm whose origin remains to be explored. It is not yet known whether this circadian rhythm extends to the neuroendocrine system. It is certainly present at 100 days of gestational age and perhaps much earlier. One may well speculate as to whether this rhythm could be the basis for the timing system which normally determines the onset of parturition in sheep with such remarkable exactitude. That it can be thrown out of gear by patho-physiological processes, such as hypoxaemia, seems very likely. But it would be unwise to conclude at the present moment that all processes which lead to premature delivery operate through the pituitary–ACTH–adrenocortical mechanism. There are other possibilities of interaction at a local level, for instance by the release of prostaglandin $F_{2\alpha}$.

It is also appropriate at this point to bring together some of the events which in the sheep appear to precede the onset of labour. Previous accounts of this process have given rise to the impression that established labour is a relatively sudden and brief affair in sheep. Yet examination of the amniotic fluid pressure, continuously recorded night and day, shows clear evidence of an irregular disturbance for many hours preceding the appearance of large regular uterine contractions. Even before this there are observations which suggest the initiation of parturition. These include a reduction in foetal breathing movements, a reduction in the volume of swallowing and a decrease in outward tracheal flow. As K. Boddy and J. Robinson (unpublished observations) have found, these events are also associated with a readily perceptible erythema of the anal and vaginal regions of the ewe which, one may speculate, could be due to a vasodilatation through a rise in plasma oestrogen concentration.

ACKNOWLEDGEMENTS

This work has been carried out with the aid of grants from the Medical Research Council. I must also thank my colleagues, already mentioned by name, who have allowed me to quote their unpublished findings, and others who helped in gathering the data, including Drs W. Harris, K. Bolton, W. Dodson, N. Green, A. Holmes, J. McCairns, H. Holt and A. Stevens.

REFERENCES

ALEXANDER, D. P., BRITTON, H. G., FORSLING, M. L., NIXON, D. A. & RATCLIFFE, J. G. (1972). Adrenocorticotrophin and vasopressin in foetal sheep and the response to stress. In *Endocrinology of Pregnancy and Parturition. Studies in Sheep*, ed. C. G. Pierrepoint, p. 112. Cardiff: Alpha Omega Alpha Publishing.

BERNHARD, G. C. & MEYERSON, B. A. (1968). Early ontogenesis of electrocortical activity. In *Clinical Electroencephalography of Children*, ed. P. Kellaway & I. Petersen, pp. 11–29. Stockholm: Almquist and Wiksell.

BODDY, K., DAWES, G. S. & ROBINSON, J. (1973). A 24-hour rhythm in the foetus. In *Proceedings of the Sir Joseph Barcroft Centenary Symposium on Foetal and Neonatal Physiology*, pp. 63–66. Cambridge University Press.

BODDY, K. & MANTELL, C. D. (1972). Observations of fetal breathing movements transmitted through maternal abdominal wall. *Lancet*, ii, 1219–1220.

BODDY, K. & ROBINSON, J. (1971). External method for detection of fetal breathing *in utero*. *Lancet* ii, 1231–1233.

BRADLEY, R. & MISTRETTA, C. (1973). The sense of taste and swallowing activity in foetal sheep. In *Proceedings of the Sir Joseph Barcroft Centenary Symposium on Foetal and Neonatal Physiology*, pp. 77–81. Cambridge University Press.

BROUGHTON PIPKIN, F., MOTT, J. C. & ROBERTON, N. R. C. (1971). Angiotensin II-like activity in circulating arterial blood in immature and adult rabbits. *J. Physiol., Lond.* **218**, 385–403.

BUDDINGH, F., PARKER, H. R., ISHIZAKI, G. & TYLER, W. S. (1971). Long term studies of the functional development of the fetal kidney in sheep. *Am. J. vet. Res.* **32**, 1993–1998.

DAWES, G. S., FOX, H. E., LEDUC, B. M., LIGGINS, G. C. & RICHARDS, R. T. (1970). Respiratory movements and paradoxical sleep in the foetal lamb. *J. Physiol., Lond.* **210**, 47–48P.

DAWES, G. S., FOX, H. E., LEDUC, B. M., LIGGINS, G. C. & RICHARDS, R. T. (1972). Respiratory movements and rapid eye movement sleep in the foetal lamb. *J. Physiol., Lond.* **220**, 119–143.

ESPINAR, E. A., HART, D. S. & BEAVEN, D. W. (1972). Cortisol secretion during acute stress and response to dexamethasone in sheep with adrenal transplants. *Endocrinology*, **90**, 1510–1514.

GILLAN, G. C., JACOBS, L. S., FRAM, D. H. & SNYDER, F. (1972). Acute effect of a glucocorticoid in normal human sleep. *Nature, Lond.* **237**, 398–399.

LIGGINS, G. C. (1969). The foetal role in the initiation of parturition in the ewe. In *Foetal Autonomy, Ciba Foundation Symposium*, eds. G. E. W. Wolstenholme & M. O'Conner. Churchill: London.

LIGGINS, G. C., GRIEVES, S. A., KENDALL, J. Z. & KNOX, B. S. (1972). The physiological roles of progesterone, oestradiol-17β and prostaglandin F$_{2\alpha}$ in the control of ovine parturition. *J. Reprod. Fert.* Suppl. **16**, 85–103.

MCNATTY, K. P., CASHMORE, M. & YOUNG, A. (1972). Diurnal variations in plasma cortisol levels in sheep. *J. Endocr.* **54**, 361–362.

MERLET, C., HOERTER, J., DEVILLENEUVE, C. & TCHOBROUTSKY, C. (1970). Mise en évidence de movements respiratoires chez le foetus d'agneau in utero au cours du dernier mois de la gestation. *C. r. hebd. Séanc. Acad. Sci., Paris.* **270**, 2462–2464.

MOTT, J. C. (1973). The renin–angiotensin system in foetal and newborn mammals. In *Proceedings of the Sir Joseph Barcroft Centenary Symposium on Foetal and Neonatal Physiology*, pp. 166–180. Cambridge University Press.

RUCKEBUSCH, Y. (1971). Activité électro-corticale chez le foetus de la brebis (*Ovis aries*) et de la vache (*Bos taurus*). *Revue Méd. vét.* **122**, 483–510.

RUCKEBUSCH, Y. (1972). Development of sleep and wakefulness in the foetal lamb. *Electroenceph. clin. Neurophysiol.* **32**, 119–128.

THORBURN, G. D., NICOL, D. H., BASSETT, J. M., SHUTT, D. A. & COX, R. I. (1972). Parturition in the goat and sheep: changes in corticosteroids, progesterone, oestrogens and prostaglandin F. *J. Reprod. Fert.* Suppl. **16**, 61–84.

TRIMPER, C. E. & LUMBERS, E. R. (1972). The renin–angiotensin system in foetal lambs. *Pflügers Arch. ges. Physiol.* **336**, 1–10.

THE ROLE OF PROGESTERONE IN HUMAN LABOUR

By LARS P. BENGTSSON

INTRODUCTION

Since I intend to concentrate on only one of the many factors involved in human labour – the action of progesterone – I wish to emphasise before I begin how hazardous it is to study only a single link in a long chain and how extremely cautious one must be when drawing conclusions from the results of experimental studies confined to a single aspect.

There is no doubt that the factors governing labour are many and complex. The effects of a single hormone are complex too and to understand them there is much that we must know. For example where and how the hormone is synthesised, what quantities are produced in different physiological states, by what means it reaches the circulation, how it is transported in the circulation and removed from it, how it is metabolized in the blood and in its target organs, what changes it undergoes in various tissues and, finally, in what manner it is excreted. Some of these questions can be answered as regards progesterone, others cannot.

Every research worker observes a problem from an individual angle. I ask your indulgence, therefore, if I concentrate on that aspect of the problem of progesterone action which is of most interest to me and to my colleagues in Lund, namely, the question whether progesterone in the pregnant woman – as in the rabbit – depresses myometrial activity and sensitivity and whether the onset of labour is due to a withdrawal of this effect. This problem can be studied in conditions where the myometrium changes from inactivity to activity as, for example, in normal pregnancy and labour and in abortion, both spontaneous and induced. Occasionally, studies on the non-pregnant uterus may also be of help. In these conditions the following parameters can be measured: (i) progesterone production, (ii) blood progesterone concentration, (iii) excretion of progesterone metabolites, (iv) progesterone concentration in the myometrium, (v) spontaneous myometrial activity, (vi) myometrial activity after stimulation and (vii) the effect of progesterone administration on spontaneous and induced myometrial activity.

NORMAL PREGNANCY AND LABOUR

We have found that in mid-pregnancy, progesterone is produced at the rate of about 75 mg/24 h, and at term at the rate of about 250 mg/24 h. In a twin pregnancy as much as 500 mg/24 h have been reported (Bengtsson & Ejarque, 1964). These values correspond well with the findings of other workers (for references see Kincl, 1971) and indicate that a tenfold increase in progesterone production takes place during pregnancy. No studies on progesterone production during labour appear to have been performed. It is of interest, however, that Haskins (1954) found less progesterone in placentas from women delivering normally than in placentas from women delivered by Caesarean section before the onset of labour. Unfortunately it is not known whether this decrease was the cause or the result of myometrial activity.

As a consequence of increased progesterone production, blood progesterone concentration also rises during human pregnancy. Most investigators (for references see Bengtsson 1971), with the exception of Csapo and his coworkers (1971), have failed to demonstrate a drop in progesterone concentration in the peripheral blood before the onset of labour. Llauró, Runnebaum & Zander (1968) found no statistically significant drop until after removal of the placenta. In a meticulous study on a woman from the 16th day before to the 6th day after delivery, Woolever & Goldfien (1965) found that the drop in blood progesterone concentration started 45 min after the onset of labour.

Before the present sensitive techniques for the accurate determination of blood progesterone concentration were available, urinary pregnanediol at term was often measured as a parameter of progesterone production. However, a significant drop in pregnanediol excretion before the onset of labour has so far not been reported (for references see Bengtsson, 1971).

The concentration of progesterone in its target tissue, the myometrium, is of great importance in a consideration of the role of progesterone in the onset of labour. Particular interest in this aspect of the action of progesterone has been engendered by Csapo's hypothesis of a local effect of progesterone on the placenta and the 'asymmetric' uterus: 'the placenta gives up its progesterone not to the blood stream, as all endocrine glands are believed to do, but by direct diffusion into the neighbouring myometrium' (Csapo, 1959). Runnebaum & Zander (1971) have carried out a detailed study of progesterone concentrations in different parts of the myometrium at different stages of human pregnancy. In early pregnancy the highest progesterone concentration was found at the placental site and the concentration decreased with increasing distance from this locus. This finding may indicate a local effect of the placenta. At later stages of pregnancy no further change was observed in the concentration of progesterone at the placental site but an increase in concentration was found at the antiplacental site. Thus the concentration of progesterone in the myometrium of the pregnant uterus changes from an asymmetric to a symmetric distribution. This finding does not support the theory that the onset of labour is the result of a withdrawal of a progesterone block. The authors conclude that differences in progesterone concentration between the placental and antiplacental regions of the myometrium may depend upon the differential distribution of progesterone binding sites.

Recently, Pulkkinen & Enkola (1972) found a gradient in progesterone concentration in the foetal membranes, the concentration being highest in the region close to the placenta. However, after rupture of the membranes and the consequent stimulation of uterine activity, no alteration in the concentration of progesterone in the membranes was found (M. O. Pulkkinen, personal communication).

There is no clear correlation between spontaneous and oxytocin-induced myometrial activity towards term and blood progesterone levels. It is of interest, however, that cases with high myometrial sensitivity to oxytocin at term more often show low progesterone concentrations in the peripheral blood (Johansson, 1968). Administration of high doses of progesterone or medroxyprogesterone do not prevent the onset or progress of labour (Brenner & Hendricks, 1962; Csapo, de Sousa-Filho, de Souza & de Souza, 1966). It must be concluded at present that very few facts indicate that the onset of normal labour at term is the result of the withdrawal of a progesterone block.

ABNORMAL PREGNANCY AND LABOUR

There can be few obstetricians nowadays who still believe that threatened or habitual abortion and threatened or habitual premature delivery are caused exclusively by a drop in progesterone activity or that they can be prevented or treated by the administration of progesterone. Double-blind studies have demonstrated the inefficiency of gestagen treatment in these conditions (Møller, Wagner & Fuchs, 1964; Kerr *et al.* 1966).

The mysterious form of abnormal pregnancy called 'missed abortion' in which a dead foetus is retained for weeks or even months was studied by our group some years ago (Bengtsson & Falk, 1964; Bengtsson & Forsgren, 1966). It has long been known that most missed abortions occur when the placenta survives after foetal death. Cassmer (1959) showed in patients admitted for therapeutic abortion that when the foetus dies *in utero*, and the placenta is left intact, urinary oestriol excretion drops very rapidly, whereas pregnanediol excretion is maintained until the conceptus is surgically removed some days later. We found the same endocrine changes in cases of spontaneous intrauterine foetal death which had developed into missed abortion: there was a rapid decrease in urinary oestriol excretion and a very slow drop in pregnanediol excretion, reaching non-pregnant values after several weeks only. It is difficult to define the specific role of progesterone in the development of missed abortion, but we believe that an abnormality in the time sequence of the decrease of oestrogen and progesterone effects may cause the myometrium's failure to contract. Our study has also resulted in an effective treatment for missed abortion.

Thus, in threatened and habitual abortion and threatened and habitual premature labour and missed abortion, a role of progesterone and progesterone withdrawal cannot be excluded but remains obscure.

ARTIFICIALLY PRODUCED ABNORMAL CONDITIONS

Artificially produced conditions, of which the most common is therapeutic abortion, have been studied by many groups. In this section I will discuss mainly studies of saline-induced abortion. About a decade ago, Bengtsson & Csapo (1962) studied the evolution of spontaneous and oxytocin-induced myometrial activity after the intra-amniotic injection of 20 per cent sodium chloride solution. The results suggested the withdrawal of a myometrial block, possibly a progesterone block. Recently, Gustavii (1971) added radioactive sodium to the saline solution and studied the spread of the radioactive label in the uterine tissues. He found that the labelled sodium had penetrated a few millimetres into the placenta, had penetrated very little into the decidua and almost not at all into the myometrium.

The question to be answered now is whether placental damage caused by saline injection is enough to depress progesterone production, thereby withdrawing the supposed progesterone block. Csapo and coworkers (1969) studied the blood progesterone concentration in six women undergoing saline-induced abortion and found a significant decrease before the onset of abortion. Nilsson & Bengtsson (1973, in preparation) studied blood progesterone concentrations in 18 women undergoing saline-induced abortion and found the same mean values as Csapo *et al.* (1969) but there was no relationship between blood progesterone concentration and the onset or outcome of abortion when the values obtained for each patient were considered individually. Similar results were reported by Holmdahl, Johansson & Nilsson (1971). Short, Wagner, Fuchs & Fuchs (1965) could find no consistent change in progesterone concentration in the uterine vein blood of women after the injection of intra-amniotic saline for the induction of abortion. Attempts to halt the evolution of myometrial activity after saline injection by the administration of a gestagen have been unsuccessful (Møller *et al.* 1964; Kerr *et al.* 1966) though other workers have reported a small, inconclusive effect (Bengtsson & Csapo, 1962; Gennser, Kullander & Lundgren, 1968). In our present state of knowledge, it does not seem likely that the onset of myometrial activity which follows the intrauterine injection of hypertonic saline is caused by a decrease in progesterone levels. I believe that we must look for other factors to explain the effect of hypertonic saline on myometrial activity.

I should like to mention here the results of some experiments on saline-induced abortion (L. P. Bengtsson and M. Martinez-Gaudio, unpublished results). In normal pregnancy close to term, the sensitivity of the myometrium to oxytocin increases to a greater extent than its sensitivity to vasopressin. If the mechanism engendering the onset of uterine contractions in saline-induced abortion were the same as that in spontaneous labour at term, one would expect this divergence in degree of myometrial sensitivity to oxytocin and vasopressin to occur before the onset of induced abortion. We found that the myometrium responded with equal and increasing sensitivity to both oxytocin and vasopressin after the saline injection and that no disproportion in degree of sensitivity developed either during or immediately after abortion. This finding supports the view that myometrial activity is controlled in diverse ways and that contractions can be initiated by changes in a variety of factors.

The role of the corpus luteum in early spontaneous abortion is a topic which has received much attention in the past. This problem has been taken up again by Csapo *et al.* (1972) who studied the effect of removal of the corpus luteum in 12 women in early pregnancy. When the operation was performed at about the seventh week of pregnancy, they observed a decrease in blood progesterone concentration, an increase in spontaneous and oxytocin-induced uterine activity and, finally, abortion after about 5 days. When, however, the removal of the corpus luteum took place in the eighth or ninth week of pregnancy, there was no fall in blood progesterone concentration, no increase in myometrial activity and no abortion. It was concluded that before the placenta takes over the production of progesterone, the corpus luteum is indispensable for the maintenance of pregnancy. The effect of progesterone and of progesterone withdrawal on the myometrium is not stressed in this report since the rapid decrease in progesterone concentration after removal of the corpus luteum was followed by abortion as much as 6 days later, which may indicate a disturbance of the life and function of the whole conceptus.

THE NON-PREGNANT STATE

I now come to the last category. I call these studies '*faute de mieux*' because in pregnancy and labour the factors controlling myometrial activity are very numerous and it is difficult to isolate a single factor and study its significance in a defined and predictable way. In the uterus of the non-pregnant woman far fewer factors are involved and it is possible to create a well-defined endocrinological situation as, for example, when studying the effects of hormone administration after the menopause or ovariectomy. It is important to remember, however, that pregnancy and non-pregnancy are widely differing states and that the pregnant and non-pregnant uterus may almost be regarded as two different organs. Nevertheless, I firmly believe that an understanding of the endocrine control of the non-pregnant uterus will make a significant contribution towards our eventual understanding of the factors controlling labour, although at our present stage of knowledge we must beware the temptation to extrapolate from the non-pregnant to the pregnant state.

Does progesterone depress myometrial activity in the uterus of the non-pregnant woman? I believe this to be a crucial question.

The different patterns of uterine activity occurring during the menstrual cycle have been frequently described. Most modern authors agree that from the end of menstrual bleeding to some days after ovulation frequent contractions of low amplitude and short duration can be recorded. From about the 18th to 20th days contractions with lower frequency, higher amplitude and longer duration are superimposed upon the smaller ones. As menstruation approaches, the latter type of contraction becomes more pronounced creating a pattern which has been described as 'pre-labour like'. During the first days of menstrual bleeding there are strong and regular contractions giving a recording which cannot be distinguished from a recording obtained during active labour. This pattern of contractions is therefore called 'labour like'. If we relate these different types of uterine contraction to the concentrations of oestrogen and progesterone in the blood at corresponding periods of the cycle, we find that the small, frequent contractions coincide with oestrogen predominance, 'pre-labour like' activity coincides with the presence of both oestrogen and progesterone, and 'labour like' activity with the withdrawal of both hormones. In conditions where oestrogen only is produced (e.g. in anovulatory cycles and the Stein–Leventhal syndrome), only the small, frequent contractions can be recorded (Moawad & Bengtsson, 1967, 1968a, b). Bleeding in anovulatory cycles in not accompanied by any change in motility pattern. In postmenopausal or ovariectomized women treated with oestrogen alone, a pattern of contractions corresponding to that of the proliferative phase was observed (Moawad & Bengtsson, 1968b). Withdrawal of oestrogen treatment did not change the pattern (Moawad, 1970). Administration of progesterone alone to postmenopausal women produced a tendency towards 'pre-labour like' activity. Progesterone or norethisterone administration in the proliferative phase resulted in 'pre-labour like' activity (Bengtsson, 1970a). When large doses of oestrogen and gestagen were given together, quite strong and regular contractions were recorded.

Can progesterone inhibit the propagation of myometrial contractions in the non-pregnant uterus? This problem must be broken down into several subsidiary questions. Are contraction waves present in the non-pregnant uterus? Is the uterine cavity a real cavity? If it is, pressure recordings taken at different levels of the uterus should give the same reading. Different pressure recordings at different levels can be obtained only if the so-called uterine cavity is not a cavity at all but merely a slit between the anterior and posterior wall of the uterus.

To clarify these uncertainties, we undertook the following investigation. In 14 healthy, parous women of fertile age, intrauterine pressure was recorded at two or three different levels by means of small rubber balloon catheters. A total of 24 cycles were studied. From the end of menstruation up to the middle of the secretory phase we found the usual small, frequent contractions. Due to the high frequency and regularity of the contractions, it was difficult to study their propagation. The contractions often looked propagated or even synchronous. A more detailed examination of the recordings, however, revealed that the frequency was different at different levels of the uterus and that the amplitude also changed independently at different levels. This indicates that at this period of the cycle there are no contraction waves but only local contractions, giving the impression of 'kneading' movements in the myometrium.

From the mid-secretory phase to menstruation we observed a gradual change from the local contractions towards the more clearly propagated ones characteristic of menstruation. During menstrual bleeding, propagation improves and real contraction waves can be discerned. Some waves move downwards, some upwards and some start in the middle and go in both directions. Completely synchronous contractions occurred but they were rare.

These findings indicate that in the non-pregnant woman progesterone does not depress myometrial activity or the propagation of contractions. The experiments just described also established that the non-pregnant human uterus contains a slit but no real cavity.

Finally, I come to a consideration of the effect of progesterone and other gestagens on myometrial sensitivity. We have studied the effect of gestagens on myometrial sensitivity to oxytocin, vasopressin and to the presence of a Lippes' loop (Bengtsson & Moawad, 1967; Bengtsson, 1970 b, c; Moawad, 1970; Moawad & Bengtsson, 1970). Oxytocin and vasopressin were administered in single intravenous injections of 0·1 or 0·2 i.u. In the proliferative phase oxytocin had no effect and vasopressin had a small stimulatory effect on the normal patterns of uterine contraction. In the secretory phase oxytocin was ineffective while vasopressin had a marked stimulatory effect. During the first days of menstruation oxytocin was again ineffective whereas vasopressin exerted a very vigorously stimulatory effect. Progesterone administered during the proliferative phase produced the same vasopressin sensitivity as is found in normal secretory phases. Uterine sensitivity to oxytocin could not be enhanced by either oestrogen or progesterone administration. In studies of menstrual cycles with the Lippes' loop *in situ* we observed a precocious evolution of 'labour like' activity in the secretory phase. From the results of these investigations we concluded that myometrial sensitivity to oxytocin, vasopressin or the Lippes' loop was not depressed by progesterone.

My final conclusion must be that while the role of progesterone in human labour remains obscure, its importance in the control of myometrial function in the pregnant woman cannot be ignored. Further studies are clearly needed and when we are able to study progesterone binding in the myometrium, and variations in binding capacity, there will be new possibilities of solving the old question of the role of progesterone in human labour.

REFERENCES

BENGTSSON, L. P. (1970 a). Studies on the effect of gestagen on myometrial *in vivo* activity in non-pregnant women. *Acta obstet. gynec. scand.* **49** (Suppl. 6), 5–12.

BENGTSSON, L. P. (1970 b). Myometrial activity in a double non-pregnant human uterus. *Acta obstet. gynec. scand.* **49** (Suppl. 6), 13–18.

BENGTSSON, L. P. (1970 c). Effect of progesterone upon the *in vivo* response of the human myometrium to oxytocin and vasopressin. *Acta obstet. gynec. scand.* **49** (Suppl. 6), 19–25.

BENGTSSON, L. P. (1971). Progesterone and its metabolites in blood and urine. In *International Encyclopedia of Pharmacology and Therapeutics*, ed. M. Tausk, Section 48, vol. 1, pp. 413–440. Oxford: Pergamon.

BENGTSSON, L. P. & CSAPO, A. (1962). Oxytocin response, withdrawal and reinforcement of defense mechanism of the human uterus at mid-pregnancy. *Am. J. Obstet. Gynec.* **83**, 1083–1093.

BENGTSSON, L. P. & EJARQUE, P. (1964). Production rate of progesterone in the last month of human pregnancy. *Acta obstet. gynec. scand.* **43**, 49–57.

BENGTSSON, L. P. & FALK, V. (1964). Missed abortion. In *Obstetrik och Gynekologi*, eds. S. Kullander & A. Sjövall, pp. 77–83. Lund, Sweden: Berlingska.

BENGTSSON, L. P. & FORSGREN, B. (1966). The diagnosis of intra-uterine foetal death and the elucidation of the aetiology of 'missed abortion' by means of semi-quantitative gas chromatographic determination of urinary oestriol and pregnanediol. *Acta obstet. gynec. scand.* **45**, 155–175.

BENGTSSON, L. P. & MOAWAD, A. (1967). The effect of Lippes' loop on human myometrial activity. *Am. J. Obstet. Gynec.* **98**, 957–965.

BRENNER, W. E. & HENDRICKS, C. H. (1962). Effect of medroxyprogesterone acetate upon the duration and characteristics of human gestation and labor. *Am. J. Obstet. Gynec.* **83**, 1094–1098.

CASSMER, O. (1959). Hormone production of the isolated human placenta. *Acta endocr., Copenh.* **32** (Suppl. 45), 1–82.

CSAPO, A. (1959). Function and regulation of the myometrium. *Ann. N.Y. Acad. Sci.*, **75**, 790–808.

CSAPO, A., KNOBIL, E., VAN DER MOLEN, H. J. & WIEST, W. G. (1971). Peripheral plasma progesterone levels during human pregnancy and labor. *Am. J. Obstet. Gynec.* **110**, 630–633.

CSAPO, A. I., KNOBIL, E., PULKKINEN, M., VAN DER MOLEN, H. J., SOMMERVILLE, I. F. & WIEST, W. G. (1969). Progesterone withdrawal during hypertonic saline-induced abortions. *Am. J. Obstet. Gynec.* **105**, 1132–1134.

CSAPO, A. I., PULKKINEN, M. O., RUTTNER, B., SAUVAGE, J. P. & WIEST, W. G. (1972). The significance of the human corpus luteum in pregnancy maintenance. I. Preliminary studies. *Am. J. Obstet. Gynec.* **112**, 1061–1067.

CSAPO, A., DE SOUSA-FILHO, M. B., DE SOUZA, J. C. & DE SOUZA, O. (1966). Effect of massive progestational hormone treatment on the parturient human uterus. *Fert. Steril.* **17**, 621–636.

GENNSER, G., KULLANDER, S. & LUNDGREN, N. (1968). Studies of therapeutic abortions induced by injection of hypertonic saline. *J. Obstet. Gynaec. Br. Commonw.* **75**, 1058–1062.

GUSTAVII, B. (1971). Studies of the abortive mechanism of intrauterine injection of hypertonic saline. *Acta obstet. gynec. scand.* **50**, 43.

HASKINS, A. L. (1954). The progesterone content of human placentas before and after the onset of labor. *Am. J. Obstet. Gynec.* **67**, 330–338.

HOLMDAHL, T. H., JOHANSSON, E. D. B. & NILSSON, B. A. (1971). Plasma progesterone in pregnancy interrupted by the intrauterine injection of hypertonic saline. *Acta endocr., Copenh.* **66**, 82–88.

JOHANSSON, E. D. B. (1968). Progesterone level and response to oxytocin at term. *Lancet*, ii, 570.

KERR, M. G., ROY, E. J., HARKNESS, R. A., SHORT, R. V. & BAIRD, D. T. (1966). Studies of the mode of action of intra-amniotic injection of hypertonic solutions in the induction of labor. *Am. J. Obstet. Gynec.* **94**, 214–224.

KINCL, F. A. (1971). Chemistry and biochemistry of progesterone. In *International Encyclopedia of Pharmacology and Therapeutics*, ed. M. Tausk, Section 48, vol. 1, pp. 13–64. Oxford: Pergamon Press.

LLAURÓ, J. L., RUNNEBAUM, B. & ZANDER, J. (1968). Progesterone in human peripheral blood before, during and after labor. *Am. J. Obstet. Gynec.* **101**, 867–873.

MOAWAD, A. (1970). The effects of estrogen withdrawal on the non-pregnant human myometrium. *Acta obstet. gynec. scand.* **49** (Suppl. 6), 27–30.

MOAWAD, A. H. & BENGTSSON, L. P. (1967). *In vivo* studies of the motility patterns of the non-pregnant human uterus. I. The normal menstrual cycle. *Am. J. Obstet. Gynec.*, **98**, 1057–1064.

MOAWAD, A. H. & BENGTSSON, L. P. (1968a). *In vivo* studies of the motility patterns of the non-pregnant human uterus. III. The effect of anovulatory pills. *Am. J. Obstet. Gynec.* **101**, 473–478.

MOAWAD, A. H. & BENGTSSON, L. P. (1968 b). *In vivo* studies of the motility patterns of the non-pregnant human uterus. II. The effect of oestrogen on human myometrial activity. *Acta obstet. gynec. scand.* **47**, 225–232.

MOAWAD, A. H. & BENGTSSON, L. P. (1970). The long term effect of Lippes' Loop in the *in vivo* motility patterns of the non-pregnant human uterus. *Acta obstet. gynec. scand.* **49** (Suppl. 6), 31–34.

MØLLER, K. J. A., WAGNER, G. & FUCHS, F. (1964). Inability of progestagens to delay abortion induced with hypertonic saline. *Am. J. Obstet. Gynec.* **90**, 694–696.

PULKKINEN, M. O. & ENKOLA, K. (1972). The progesterone gradient of the human fetal membranes. *Int. J. Gynec. Obstet.* **10**, 93–94.

RUNNEBAUM, B. & ZANDER, J. (1971) Progesterone and 20α-dihydroprogesterone in human myometrium during pregnancy. *Acta endocr., Copenh.* **66** (Suppl. 150), 5–50.

SHORT, R. V., WAGNER, G., FUCHS, A.-R. & FUCHS, F. (1965). Progesterone concentrations in uterine venous blood after intra-amniotic injection of hypertonic saline in midpregnancy. *Am. J. Obstet. Gynec.* **91**, 132–136.

WOOLEVER, C. A. & GOLDFIEN, A. (1965). Serial studies of plasma progesterone by a double isotope derivative technique. In *Hormonal Steroids*, eds. L. Martin & A. Pecile, vol. 2, pp. 253–262. New York: Academic Press.

THE ROLE OF OESTROGENS IN THE
ONSET OF LABOUR

By ARNOLD KLOPPER

One in whom persuasion and belief
Had ripened into faith, and faith become
A passionate intuition.

William Wordsworth

That certain hormones such as the oestrogens or progesterone played a central part in the onset of labour, was a hypothesis first tentatively advanced soon after these steroids were discovered. Some 40 years later it had become an article of faith among obstetricians and, to judge by the intemperate language of the more fervid advocates of one or other theory, a matter of passionate intuition. It is my objective to examine, as far as oestrogens are concerned, the grounds on which this belief is based. There is a body of evidence to suggest that the mechanism controlling the onset of labour may differ from one species to another. This review will relate to labour in women.

Although reports of normal and abnormal labour have accumulated in the literature, experimentation as such is almost impossible and it becomes necessary to rely on the findings in other species. This applies particularly to the endocrine changes which may take place in the foetus before labour. Using animals, surgical techniques have been devised by which it is possible to implant catheters in various parts of the foetal circulation while the foetus is still in the uterus and thus measure changes in hormone concentration which take place in the foetal circulation as labour approaches. The view that labour starts in response to a stimulus arising from the foetus itself, has much to recommend it. When first advanced, this hypothesis was mainly based on experimental findings in sheep (Liggins, Kennedy & Holm, 1967). It was found that if the foetal lamb *in utero* had its pituitary destroyed or adrenals removed, the ewe did not go into labour at term. It seemed possible therefore that labour was initiated by some secretion produced as a result of pituitary–adrenal activity; a suggestion supported by Liggins' (1968) finding that infusion of cortisol or corticotrophin into the foetal lamb leads to premature labour. It then remained to show that under normal circumstances the onset of labour in the ewe is preceded by an increase in foetal corticosteroid levels.

[47]

This was demonstrated by Bassett & Thorburn (1969) and indeed the mechanism whereby the pituitary gland of the foetal lamb and goat initiates labour has now been described in convincing detail, as can be seen in the contributions to this Memoir from the laboratories of G. C. Liggins and G. D. Thorburn.

The number of different oestrogens which have been isolated from the urine of pregnant women is a matter of somewhat academic debate; 20 compounds are present at least. Some of these are the products of maternal metabolism, but clearly a large variety of oestrogens are passing from the foeto-placental unit to the mother. It is an unproven, but very likely, assumption that oestrogens act in their initial passage through the myo-metrium from the foeto-placental circulation to the maternal circulation, rather than on the subsequent passage of maternal arterial blood through the uterus. Indeed at this point the major weakness of all experimental examinations of the role of oestrogens in the onset of human labour becomes apparent. Such data as are available pertain to the oestrogens to be found in the peripheral circulation and urine of the mother and not to those present at the level of the myometrium. Nevertheless, the problem of which particular oestrogen or oestrogens, among some 20 or so, to consider first remains. Some investigators have tried to solve the problem by group estimations such as measuring 'total' immunoreactive oestrogens in blood. Such 'total' oestrogen measurements are valid only if they represent a simple arithmetic sum in which each of the component parts has an equal value – a demonstrably false proposition. The alternative is an examination of the role of individual oestrogens in the onset of labour. This approach again poses the question of which particular oestrogen is most likely to play a leading role in the onset of labour in women for there is evidence to suggest that different oestrogens may be involved in different species. Thus in the sheep the principal oestrogen is oestradiol-17β while in the goat it is oestrone or oestradiol-17α (Thorburn *et al.* 1972), and in the horse equilenins are probably involved. In women two oestrogens can be singled out as possible candidates: oestriol and oestradiol-17β. Oestriol is present in the urine of non-pregnant women to a greater extent than any other oestrogen and during the course of pregnancy its urinary excretion increases by some 1000- to 10 000-fold. Although its occurrence in animals, and indeed in plants, is widespread, its role appears to be that of a rather inert endpoint in oestrogen metabolism. Only in human pregnancy is there reason to suppose, at least on grounds of quantity, that it is of importance in its own right. Although oestriol is a minor component of the oestrogens in the urine of most pregnant animals, and by far the largest fraction of the oestrogens present in the urine of pregnant women, urinary output may be deceptive, for in the peripheral circulation of the pregnant woman the concentration of unconjugated oestradiol-17β may be greater than that of oestriol (Tulchinsky, Hobel, Yeager & Marshall, 1972). There is insufficient evidence to decide whether oestradiol-17β or oestriol is the compound which exerts the effect of oestrogens on the myometrium. In this presentation I will concentrate on the role of oestriol because the foeto-placental unit synthesises more of this steroid than any other oestrogen; a metabolic peculiarity of human pregnancy. It may be that my conclusions reflect only the choice of the wrong oestrogen and are not indicative of the function in pregnancy of oestrogens as a whole.

A simplistic interpretation of the role of oestrogens in the onset of labour sets out to provide an interpretation in terms of their direct action on target organs and ignores the role of other hormones such as progesterone, oxytocin and prostaglandins in the process. A more modern view is to regard all these substances as having an integrated action; as individual links in a complicated chain of events. When, elsewhere in this volume, the views of Liggins and of Thorburn are considered, it is clear that such an integrated theory can be constructed. A free interpretation of the findings of these workers would suggest that the onset of labour is preceded by a rise in oestrogen production by the foeto-placental unit, a fall in progesterone production, an increase in myometrial sensitivity to oxytocin, and an increase in prostaglandin secretion from the endometrium. Which occurs first, oestrogen rise or progesterone fall (and one may be the cause of the other), is not clear. It is probable that either or both of these events precede and are the cause of the increased oxytocin sensitivity of the myometrium and of prostaglandin release. The evidence supports this view of the integrated action of hormones in parturition in the sheep. It remains to be seen how far the same model can be applied to human pregnancy.

Two experimental approaches can be used to explore the role of oestrogens in the onset of human labour. One is to examine what changes in oestrogen concentration are associated with the spontaneous onset of labour; the other is to attempt to manipulate oestrogen levels by the administration of oestrogens and to observe what effect this has on the onset of labour. Both methods of approach have been applied to human pregnancy.

CHANGES IN ENDOGENOUS OESTROGEN CONCENTRATIONS ASSOCIATED WITH THE ONSET OF LABOUR

The output of oestriol in urine rises steadily throughout pregnancy. The excretion increases briskly in the last six weeks, so that the curve of urinary oestriol turns sharply upward at 32–34 weeks. This upward inflection is variable in its time of onset and in degree so that the abrupt change is not obvious when the mean curve of a number of subjects is examined. Figure 1 shows the weekly oestriol output in a single subject and illustrates the phenomenon admirably.

The data on which this figure is based were obtained more than a decade ago. The problem posed has not been solved. Klopper & Billewicz (1963) speculated that the increase in oestriol excretion observed represented a

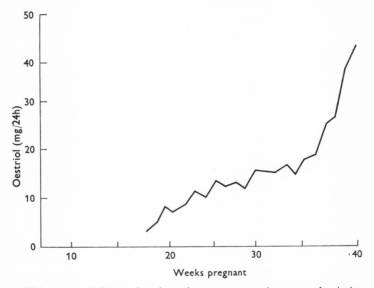

Fig. 1. Urinary oestriol excretion throughout pregnancy in a normal primigravida.

new, specifically foetal element in oestriol biosynthesis; a suggestion which is supported by the later evidence that labour starts as the result of a stimulus from the foetus.

There is other evidence which suggests that urinary oestriol levels may be linked with the onset of labour. When analysing the data from a cohort of women whose urinary oestriol output had been measured weekly throughout pregnancy, Klopper & Billewicz (1963) noted that those women who, at 36 weeks, had high oestriol concentrations, tended to go into labour at 38 or 39 weeks, while those who had low concentrations went on to 41 or 42 weeks.

The changes in urine concentration are gradual. If they are associated with the onset of labour, it must be a slow physiological process spread over the last few weeks of pregnancy, not a sudden, urgent biological signal, which precipitates labour. A consistent change immediately preceding labour has often been sought but has never been demonstrated. If such a signal exists, urinary assays do not demonstrate it. Firstly, urinary oestriol excretion in the same subject over a period is very variable. Under the most rigorously controlled conditions the 24 h urinary value has a coefficient of variation above 20 per cent (Klopper, Wilson & Cooke, 1969). If the change in oestriol production were small or short-lived it would be hidden by this high natural variability. Secondly, many metabolic processes are interposed between oestriol production in the foeto-placental unit and its ultimate excretion in the mother's urine. A large change at the myometrial level could easily be obscured by these subsequent metabolic processes.

It has long been obvious that the measurement of oestrogens in blood is a more promising line of investigation than urinary assays as a means of examining the role of oestrogens in the onset of labour. Although the results of the first assays of oestrogen in blood were published more than 30 years ago, the methods used were inadequate until recently when competitive protein binding and radioimmunoassay techniques were introduced.

These sophisticated assay methods were first applied to oestrogen measurements before labour, not in man but in domestic animals. It was found that in sheep the concentration of oestradiol-17β and of oestrone in peripheral blood started to rise very sharply in the last 48 h before the onset of labour, reaching a peak with the delivery of the first foetus (Bedford, Challis, Harrison & Heap, 1972). It seemed likely that the oestrogen being measured derived from the foeto-placental unit, for the changes were more pronounced in the uterine venous blood than in peripheral blood. Similar changes in oestrone and in oestradiol-17α concentrations take place in the blood of goats a day or two before the onset of labour (Thorburn et al. 1972) and subsequent observations, using catheters chronically implanted into the foetal circulation, showed that the spurt of oestrogen production, before labour is initiated, starts in the foetal circulation.

How far are the events in sheep and goats representative of those in man? This question was asked long before the means to the answer were available. Now that the methodological problem has to some extent been solved, a second barrier of difficulties is revealed. The fact that the pregnant woman produces a score or more oestrogens has been touched upon: the problem is compounded by the fact that each of these exists as a free or as a conjugated steroid with either or both glucosiduronic and/or sulphuric acid radicles attached at one or several sites in the molecule. In turn either the conjugated or the unconjugated steroid may be present in the plasma bound with varying firmness to a variety of proteins or in free solution. It is small wonder then that very few of this bewildering variety of oestrogens have been examined and that the choice of which compound to measure has largely been one of methodological accident in most laboratories.

Measurements of blood oestrogens in relation to the onset of labour in women have been confined to oestriol and oestradiol-17β with a few observations on oestrone. Within the circle of these three compounds investigators have chosen to measure either the unconjugated or the conjugated steroid. The motive for measuring unconjugated oestrogen derives from the assumption that this is the physiologically active moiety. It is perhaps equally likely that only that fraction of the unconjugated oestrogen not bound to protein is physiologically active, but this sub-fraction is never singled out for specific measurement, although the methodological problems associated with such assays are trivial.

Tulchinsky and his colleagues (1972) have measured unconjugated oestrone, oestradiol-17β and oestriol in plasma throughout the second half of pregnancy. Although the concentration of oestradiol-17β and of oestriol in blood rose sharply in the last few weeks of pregnancy (as had been noted for urinary excretion ten years previously) these workers failed to find the sudden spurt in the last 48 h which had been reported in sheep and goats. Their experiments were not designed to examine the oestrogen levels of the last day or two before labour specifically and in view of the finding of Sybulski (1971) that unconjugated oestradiol rose from 1·96 μg/100 ml in late pregnancy to 2·49 μg/100 ml when patients were in labour, the possibility remained that there could be undetected transient peaks of plasma oestrogens just before labour. This possibility was removed as far as unconjugated oestriol and oestradiol were concerned by the experiments of Shaaban & Klopper (1973) who measured the unconjugated plasma oestriol and oestradiol daily during the last three weeks of pregnancy and found no consistent change before the spontaneous onset of labour.

Unconjugated oestrogens probably have a fleeting existence in peripheral blood. Their concentration in the antecubital vein is as likely to be determined by their rate of metabolism by the liver as it is to be a reflection of the rate of oestrogen production by the foeto-placental unit, and may bear little relationship to the oestrogen concentration at the myometrial level. The clear demonstration of a pre-labour oestrogen surge in sheep and goats depended on measurements in uterine vein blood. If it were possible to measure oestrogen concentrations in serial samples of uterine vein blood before the onset of labour in women, equally dramatic changes might be found. The hypothesis depends on a surge of oestrogen arising from the foeto-placental unit and it is here that the answer must be sought. The first indications are not encouraging. Shearman, Jools & Smith (1972) found no difference between the concentrations of plasma oestriol of babies delivered by Caesarean section before labour and those delivered *per vaginam*. This line of investigation has not been pursued very far and until more data are available it is perhaps best regarded as an open question.

In late pregnancy very nearly all the oestriol in the maternal circulation is derived from the foeto-placental unit. Only about 10 per cent of the total oestriol in maternal plasma is present in the unconjugated form and it is likely that the remaining 90 per cent of conjugated oestriol is a more

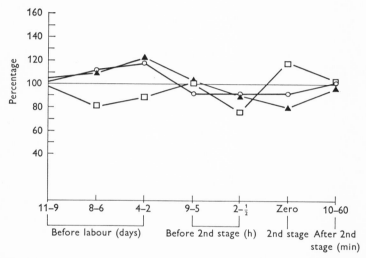

Fig. 2. The percentage change in plasma concentration of unconjugated oestriol (▲), oestriol glucosiduronate (○) and oestriol sulphate (□) before and during labour. Zero time indicates fully-established second stage labour. The results are overall mean values of five patients. (Reproduced with permission from Masson & Klopper, 1972.)

accurate indication of the amount of oestriol being produced by the foeto-placental unit than the small, variable unconjugated subfraction. Some investigators have measured total plasma oestriol on the grounds that this gives a more reliable picture of oestriol production in the foeto-placental unit. There is, however no indication from the findings of Nachtigall, Bassett, Hogsander & Levitz (1966), who measured total oestriol concentration in maternal plasma, that a surge of oestriol production immediately precedes the onset of labour.

Other workers have examined the possibility that one or other of the individual oestriol moieties in the maternal circulation may be associated with the onset of labour. Masson & Klopper (1972) measured the concentrations of unconjugated oestriol, of oestriol glucosiduronate and of oestriol sulphate before and during labour. Their findings are summarised in Fig. 2, and show that there is no marked change in any of these fractions associated with the onset of labour.

A further possibility has been mooted from time to time. This is that it may not be the absolute concentration of any particular oestrogen which controls myometrial activity but that the concentration of one or other oestrogen relative to the progesterone concentration determines myometrial contractility. Certainly the production of progesterone, as measured by urinary pregnanediol, either levels off or actually falls at term, while oestrogen production, as manifested by urinary oestriol excretion, continues to rise briskly. Thus the last few weeks of pregnancy are marked by a sharp rise in the ratio in favour of oestriol. Examination of the ratio of unconjugated oestrogens to progesterone in the maternal plasma as labour approaches has failed to establish the thesis that the onset of labour is marked by a relative progesterone dominance. Both Tulchinsky *et al.* (1972) and Shaaban & Klopper (1973) measured the concentrations of unconjugated oestradiol-17β, of unconjugated oestriol and of progesterone in plasma in late pregnancy but failed to find any critical oestrogen: progesterone ratio at which labour was precipitated.

THE EFFECT OF ADMINISTERED OESTROGEN ON THE ONSET OF LABOUR

The belief that the administration of oestrogen would influence either the onset or the course of labour took hold soon after oestrogenic preparations became available for clinical use. Techniques for the objective assessment of drug action such as blind trials against a placebo were yet far off, and the supposed effect of oestrogen on myometrial contractility became a matter of passionate intuition in obstetrical practice all over the world.

The clinical impressions of the early observers tended to establish academic reputations rather than the facts of the effects of oestrogen upon the uterus. A more objective approach appeared with the introduction of measurements of the frequency, strength and duration of uterine contractions. But measurements of uterine contractions are not, by themselves, an adequate description of labour and must be supplemented by other parameters such as induction–delivery interval or the length of labour. Here lies a snag too often ignored in clinical investigations. Measurements such as induction–delivery interval may be influenced by the effect of oestrogens on myometrial activity but they are also much affected by other factors such as the state of the cervix or cephalo-pelvic disproportion. It is a *sine qua non* that oestrogen-treated subjects and controls must be accurately matched for parity, age, cervical state, stage of gestation and maternal size. Even so, foetal size, which can hardly be anticipated, may make a critical difference to the length of labour.

A failure to standardise the clinical material may have influenced the findings of the Buenos Aires group who did much of the early work on the effect of oestrogen administration on the onset of labour (Pinto, Leon, Mazzocco & Scassera, 1967). These workers used a double-blind design, giving an intravenous infusion of either oestradiol-17β or the solvent vehicle to equal numbers of women at term. They claimed that those given the oestrogen tended to go into spontaneous labour more rapidly and to respond more readily to intravenous oxytocin. Other work from this group has emphasised the fact that the action of oestradiol is not only directly on myometrial contractility but that it also causes changes in the cervix and an enhanced sensitivity of the myometrium to oxytocin (Pinto, Fisch, Schwarz & Montuori, 1964; Pinto, Rabow & Votta, 1965). In assessing Pinto's work, account should be taken of the fact that positive effects were recorded on parameters other than induction–delivery interval. Whether or not oestrogen administration will facilitate the onset of labour must be regarded as an open question. Some investigators have noted positive findings (Järvinen, Luukkainen & Väistö, 1965) while others have failed to find any effect (Klopper & Dennis, 1962; Agüero & Aure, 1971).

Oestrogens can affect myometrial activity in two ways; two interrelated possibilities which are best examined separately. They may influence the nature of the contractions associated with labour once these are elicited by some trigger stimulus or they may themselves provide the trigger impulse. Klopper & Dennis (1962) explored the first possibility when they carried out a blind trial, giving oestriol, stilboestrol or a placebo to a group of women who were having labour induced by artificial rupture of the membranes. Their findings are summarised in Fig. 3. The criterion which they used was the induction–delivery interval. It can be seen from Fig. 3 that most patients delivered 6–12 h after the rupture of membranes, whether they were given stilboestrol, oestriol or the placebo. Indeed there was a tendency for the induction–delivery interval to be longer in the placebo-treated group though this was not statistically significant.

When oestrogens are given orally or injected into a peripheral vein, oestrogen-metabolising tissues such as the liver are interposed between the administered steroid hormone and its target cells in the uterus. It seems doubtful whether such an experimental model really approaches the

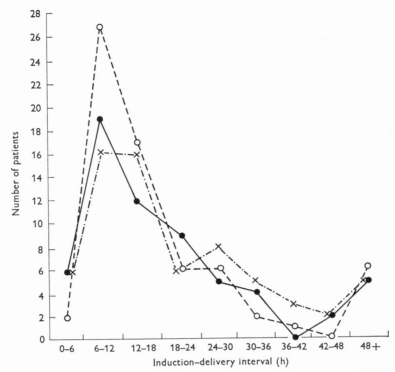

Fig. 3. Induction–delivery interval in women treated with oestriol (●), stilboestrol (○) or placebo (×) tablets before artificial rupture of the membranes. (Reproduced with permission from Klopper & Dennis, 1962.)

natural situation where the placenta is in effect conveying directly to the myometrium a constant oestrogen transfusion and where the concentration of oestrogen to which the myometrium is exposed can be altered very rapidly. The major oestrogen in the foetal circulation is oestriol-3-sulphate. In an attempt to approach the natural situation more closely, Klopper, Dennis & Farr (1969) injected large amounts of oestriol sulphate into the liquor amnii after rupture of the membranes and observed the effect of such treatment on the subsequent response of the uterus to amniotomy. They failed to demonstrate any significant difference either in uterine contractions or in induction–delivery interval between patients treated with oestriol sulphate and those who received intra-uterine injection of solvent vehicle only.

These experiments, although they suggest that an increase in oestrogen concentration at the myometrial level does not alter the response of the myometrium to an activating stimulus such as rupture of the membranes, do not solve the problem of whether increased oestrogen concentration could of itself provoke uterine action. This possibility was examined in a further therapeutic trial by the Aberdeen group (Klopper, Farr & Dennis, 1973). After recording uterine activity for 2 h in a group of primigravidae at term, amniocentesis was performed and either oestriol sulphate or the solvent injected into the liquor amnii by the transabdominal route. The effect of this injection on uterine activity was observed for 4 h in order to study the effect of the treatment on the subsequent labour. Rupture of the membranes was performed at the end of this period. The results are shown in Fig. 4. Once again the investigation failed to show any effect of intra-amniotic oestriol sulphate either on intrinsic myometrial activity or on the subsequent responses of the myometrium to the stimulus of rupture of the membranes.

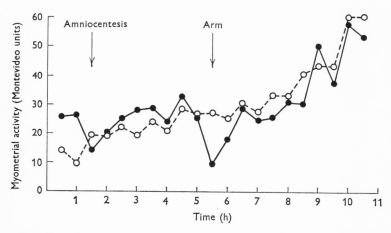

Fig. 4. The effect of the intra-amniotic injection of oestriol sulphate ($\bigcirc$) or placebo ($\bullet$) on myometrial activity at term. Ten women received oestriol; ten, placebo. Arm = artificial rupture of membranes. (Reproduced with permission from Klopper, Farr & Dennis, 1973.)

The negative results of experiments designed to influence the onset of labour by the administration of oestrogens can be questioned in four respects. Firstly, the experiments have aimed to bring about changes in myometrial contractility, whereas the main effect of oestrogens may be on the cervix. Secondly, they have been short-term experiments lasting 2 or 3 days at most, whereas the action of oestrogen may take longer to affect the uterus. Thirdly, the studies of the Aberdeen group have been devoted mainly to oestriol and it is possible that more potent oestrogens, such as oestradiol-17β, are involved. Finally, only direct uterine effects have been considered, and the other hormones involved in the onset of labour such as oxytocin, prostaglandins or progesterone have not been examined.

CONCLUSION

It may never be possible to bring the problem of the role of oestrogens in the onset of labour to a Euclidean conclusion and write 'quod erat demonstrandum'. At this stage the interim verdict has to be couched in cautious language. There are too many findings which point to some involvement of oestrogens to be ignored. On the other hand a rise of oestrogen level is not always a dominant or indeed an essential feature of the onset of labour. It may be that oestrogens are one link in a chain of endocrine events which lead to labour. At least in man there are probably routes which can by-pass the oestrogen control point, and possibly the whole series of endocrine events play only a permissive role.

ACKNOWLEDGEMENTS

The experimental work on the role of oestrogens in labour was done in collaboration with Professor John Dennis, Drs Valerie Farr, Gordon Masson, Mamdouh Shaaban and George Wilson. I am grateful for their help and to the Medical Research Council for financial support.

REFERENCES

AGÜERO, O. & AURE, M. (1971). Inutilidad de los estrogenos en la maduracion del cuello y en la induccion del parto. Ginec. Obstet. Méx. 30, 21–29.

BASSETT, J. M. & THORBURN, G. D. (1969). Foetal plasma corticosteroids and the initiation of parturition in sheep. J. Endocr. 44, 285–286.

BEDFORD, C. A., CHALLIS, J. R., HARRISON, F. A. & HEAP, R. B. (1972). The role of oestrogens and progesterone in the onset of parturition in various species. J. Reprod. Fert. Suppl. 16, 1–23.

JÄRVINEN, P. A., LUUKKAINEN, T. & VÄISTÖ, E. (1965). The effect of oestrogen treatment on myometrial activity in late pregnancy. Acta obstet. gynec. scand. 44, 258–264.

KLOPPER, A. & BILLEWICZ, W. (1963). Urinary excretion of oestriol and pregnanediol during pregnancy. J. Obstet. Gynaec. Br. Commonw. 70, 1024–1029.

KLOPPER, A. & DENNIS, K. J. (1962). Effect of oestrogens on myometrial contractions. Br. med. J. ii, 1157–1159.

KLOPPER, A., DENNIS, K. J. & FARR, V. (1969). The effect of intra-amniotic oestriol sulphate on uterine contractions. Br. med. J. ii, 786–789.

KLOPPER, A., FARR, V. & DENNIS, K. J. (1973). The effect of intra-amniotic oestriol sulphate on uterine contractility at term. J. Obstet. Gynaec. Br. Commw. 80, 34–40.

KLOPPER, A., WILSON, G. R. & COOKE, I. (1969). Studies on the variability of urinary oestriol and pregnanediol during pregnancy. J. Endocr. 43, 295–300.

LIGGINS, G. C. (1968). Premature parturition after infusion of corticotrophin or cortisol into foetal lambs. J. Endocr. 42, 323–329.

LIGGINS, G. C., KENNEDY, D. C. & HOLM, L. W. (1967). Failure of initiation of parturition after electrocoagulation of the pituitary of the foetal lamb. Am. J. Obstet. Gynec. 98, 1080–1086.

MASSON, G. M. & KLOPPER, A. (1972). Changes in plasma oestriol concentration associated with the onset of labour. *J. Obstet. Gynaec. Br. Commonw.* **79**, 970–975.

NACHTIGALL, L., BASSETT, M., HOGSANDER, V. & LEVITZ, M. (1966). Plasma estriol levels in normal and abnormal pregnancies: An index of fetal welfare. *Am. J. Obstet. Gynec.* **101**, 638–648.

PINTO, R. M., FISCH, L., SCHWARZ, R. L. & MONTUORI, E. (1964). Action of estradiol-17β upon uterine contractility and the milk-ejecting effect in pregnant women. *Am. J. Obstet. Gynec.* **90**, 99–107.

PINTO, R. M., LEON, C., MAZZOCCO, N. & SCASSERA, V. (1967). Action of estradiol-17β at term and at onset of labor. *Am. J. Obstet. Gynec.* **98**, 540–546.

PINTO, R. M., RABOW, W. & VOTTA, R. A. (1965). Uterine cervix ripening in term pregnancy due to the action of estradiol-17β. *Am. J. Obstet. Gynec.* **92**, 319–324.

SHAABAN, M. & KLOPPER, A. (1973). Changes in unconjugated oestrogens and progesterone concentration in plasma at the approach of labour. *J. Obstet. Gynaec. Br. Commonw.* **80**, 210–217.

SHEARMAN, R., JOOLS, N. & SMITH, I. (1972). Maternal and fetal venous plasma steroids in relation to parturition. *J. Obstet. Gynaec. Br. Commonw.* **79**, 212–215.

SYBULSKI, S. (1971). Determination of free estradiol-17β levels in pregnancy plasma by a competitive protein binding method. *Am. J. Obstet. Gynec.* **110**, 304–308.

THORBURN, G. D., NICOL, D. H., BASSETT, J. M., SHUTT, D. A. & COX, R. I. (1972). Parturition in the goat and sheep: changes in corticosteroids, progesterone, oestrogens and prostaglandin F. *J. Reprod. Fert.* Suppl. **16**, 61–84.

TULCHINSKY, D., HOBEL, C. J., YEAGER, E. & MARSHALL, J. R. (1972). Plasma estrone, estradiol, estriol, progesterone and 17-hydroxyprogesterone in human pregnancy. *Am. J. Obstet. Gynec.* **112**, 1095–1100.

THE POSTERIOR PITUITARY AND
THE INDUCTION OF LABOUR

By T. CHARD

INTRODUCTION

Early work on the role of the posterior pituitary in the initiation and maintenance of labour, and in particular on the place of oxytocin, was vitiated by dispute as to the validity of the assays used for measuring the circulating levels of this hormone (see Theobald, 1968; Chard, 1972). At the same time, indirect approaches to the problem, while both elegant and of considerable value, could provide no final answer as to the nature and extent of pituitary activation during spontaneous labour (Fuchs, 1971; Caldeyro-Barcia, Melander & Coch, 1971). However, the recent development of highly specific and sensitive radioimmunoassays for oxytocin, vasopressin and neurophysin has permitted a new attack on the problem. This approach has yielded the following conclusions: (1) in parturition in non-primates, there is a release of oxytocin from the maternal pituitary, maximal at the time of delivery (Chard *et al.* 1970); (2) in man, there is 'spurt' release of oxytocin from the maternal pituitary, the frequency of the spurts increasing as labour progresses (Chard, Hudson, Edwards & Boyd, 1971; Gibbens, Boyd & Chard, 1972); (3) there is a release of both oxytocin and vasopressin from the human foetal pituitary during labour (Chard *et al.* 1971); (4) there is activation of the posterior pituitary of the foetal guinea-pig at the time of delivery (Burton & Forsling, 1972). The purpose of the present paper is to consider the evidence for these conceptions, especially in the light of more recent findings by our own and other groups.

THE MEASUREMENT OF OXYTOCIN AND
VASOPRESSIN BY RADIOIMMUNOASSAY

Radioimmunoassays for oxytocin and vasopressin have been the subject of detailed investigations in both our own and other laboratories, and their advantages in terms of specificity, reproducibility, and ease of performance have been emphasised on many occasions. Comparison with the classical biological assays, on which most existing concepts are based, has almost invariably shown an excellent correlation between the two

[61]

methods of assay (e.g. McNeilly, Forsling & Chard 1971; Burton, Forsling & Martin, 1972). Dissociation of biological and immunological activity may occur, but only under experimental conditions which are unlikely to apply to endogenous release (Forsling, Boyd & Chard, 1971; James, Chard & Forsling, 1971). With detailed attention to the exclusion of non-specific effects (Chard, 1971), it may be inferred that a positive result from a radioimmunoassay represents either intact hormone, or minor variants. Nevertheless, the possibility that endogenous hormone in plasma may be in some way inaccessible to the assay (for instance if it were present in bound or precursor form) must be taken into account (Chard, 1972). Under these circumstances a biological assay, dependent as it is on the reaction of a target organ which presumably has the mechanisms for releasing active hormone will yield a positive result where the radio-immunoassay is negative. On the present evidence, such a possibility can be neither proved nor disproved.

RELEASE OF MATERNAL OXYTOCIN DURING ANIMAL PARTURITION

Direct evidence

The data obtained by radioimmunoassay (Chard et al. 1970; McNeilly et al. 1971; Allen, Chard & Forsling, 1973) have confirmed earlier obser-vations using biological assays, that there is a massive release of oxytocin at the time of delivery, but relatively little during the earlier stages of labour (for references see Chard, 1972). Any remaining doubts as to the specificity of these findings have been removed by the observation, in the goat, of a release of neurophysin simultaneous with that of oxytocin (Fig. 1) (McNeilly, Martin, Chard & Hart, 1972). Most of the information comes from studies on larger animal species (cow, goat, horse, sheep and rabbit), but recent data indicate a similar situation in the guinea-pig (Burton & Forsling, 1972). In addition, studies on the pituitary content of oxytocin in the rat (Fuchs & Saito, 1971) and the guinea-pig (Burton & Forsling, 1972) show a fall at the time of delivery. In the rat there is also a fall in the pituitary content of vasopressin; release of vasopressin during the expulsive phase of labour has been shown in the horse (Allen et al. 1973) but not in the goat (McNeilly et al. 1971).

Indirect evidence

There is indirect evidence for the involvement of the maternal pituitary in parturition. Ablation, providing that it involves the hypothalamus as

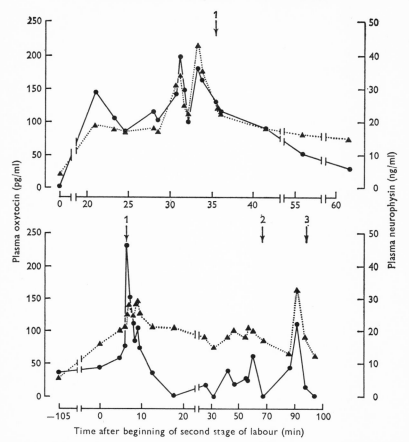

Fig. 1. The simultaneous release of oxytocin (●) and neurophysin (▲) during parturition in two goats measured in jugular vein blood. Each arrow indicates the delivery of a kid. Printed, with permission, from McNeilly *et al.* (1972).

well as the pituitary gland, can delay parturition (Fisher, Magoun & Ranson, 1938; Nibbelink, 1961), though the effect is not invariable (Gale & McCann, 1961). Administration of ethanol, a known depressant of posterior pituitary activity, will also delay labour, though there is some argument as to whether or not it has a direct effect on the uterus at the same time (see Fuchs, 1971).

The cause of oxytocin release

Our understanding of the cause of oxytocin release has advanced little since the classic work of Ferguson (1941). The mechanism involves neural reflexes from the lower genital tract and interruption of the pathway by spinal transection leads to delay in delivery (Beyer & Mena, 1970). Since the release is maximal during the expulsive phase of labour, it would seem likely that the vagina rather than the cervix is the major source of the impulses; there is excellent direct evidence in the sheep that vaginal distension leads to release of oxytocin (Roberts & Share, 1968) and that the extent of this release can be modified by the relative concentrations of circulating oestrogens and progesterone. That the presence of a foetus is unnecessary for this release is demonstrated by experiments in the horse: removal of the foetus *per abdomen* is followed, within a few hours, by delivery of the placenta *per vaginam* accompanied by release of oxytocin (Allen *et al.* 1973).

The significance of maternal oxytocin release in animals

If the fact of a release of oxytocin, and sometimes vasopressin, at the time of delivery in animals is now beyond argument, the significance of this release remains in considerable doubt. Since the levels are low in the early stages of labour, it is unlikely that an increase in oxytocin concentration plays an important part in the initiation of labour. However, it may, as suggested by Theobald (1968), play a permissive role, a low and steady level being essential for the uterine activity induced by other factors. By contrast, the abrupt release of oxytocin in the expulsive phase suggests some specific function at that time. One possibility is that it is responsible for the orderly delivery of a series of foetuses in a multiple pregnancy; such a function would render it of less importance in the larger species in which the number of young is usually small. Another possibility is that it is vital to the retraction of the uterus once emptied of its contents, thus acting as a natural haemostatic. However, such an action, occurring after the delivery of one foetus, could hardly be favourable to the environment of those which remain.

It should be noted that most studies of oxytocin release during animal parturition have measured concentrations of oxytocin in jugular venous plasma; since this represents the direct drainage of the pituitary gland, the measurements obtained are likely to be an over-estimate of the concentrations reaching the target organ. Experiments relating central and peripheral levels of endogenous hormone would be of great value.

RELEASE OF MATERNAL OXYTOCIN DURING HUMAN PARTURITION

Direct evidence

The data obtained by our group using a radioimmunoassay for oxytocin conflict with most of the earlier measurements of this hormone (see Chard, 1972). The difference is both quantitative and qualitative. Thus, the highest levels found at any stage of labour are an order of magnitude less than those reported by other workers. More important, serial sampling reveals that oxytocin release occurs in a series of spurts (Gibbens *et al.* 1972); for instance, in a sequence of eight samples collected during the first stage of labour, three may contain detectable oxytocin, while the remainder give negative results (i.e. less than 1 μu./ml in this assay). The

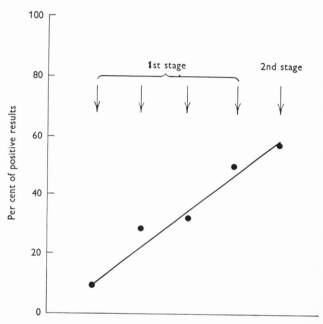

Fig. 2. The frequency of positive results (oxytocin level greater than 1μu./ml) in serial samples collected from 33 women during spontaneous labour.

frequency of 'spurt' release increases as labour progresses, reaching a maximum in the second stage (Fig. 2). There appears to be no relationship to individual uterine contractions, nor to the degree and timing of sedation. The frequency of spurts during late pregnancy and the third stage of labour has not, as yet, been studied.

The difference between these and previous results can probably be attributed to methodological factors, and it is notable that several recent studies would tend to support our findings (see F. Fuchs, this Symposium). Most earlier authors were concerned to point out that they measured 'oxytocin-like' activity, rather than oxytocin itself, reflecting the problems of specificity which arise with any biological assay for oxytocin. In theory, this difficulty should not apply to immunoassay systems which are capable of a high degree of specificity, yet the results obtained are still disparate from our own (e.g. Glick, Kumaresan, Kagan & Wheeler, 1969; Kumaresan, Anandarangam & Vasicka, 1972). A possible explanation arises from the exact method of immunoassay: our own results are derived from plasma subjected to extraction, a step designed both to concentrate the hormone, and more important, to eliminate the enzyme oxytocinase (Chard *et al.* 1970; Chard, 1971). This enzyme appears to have little effect on the handling of the hormone *in vivo*, but can have dramatic effects on a radio-immunoassay system *in vitro*. Labelled peptides, including ^{125}I-labelled oxytocin (James *et al.* 1971) are especially sensitive to its action; destruction of the label during incubation will produce a reduction in antibody-binding indistinguishable from that produced by unlabelled hormone, and thus lead to artefactual results. The effect can be produced by dilutions of plasma as high as 1 in 50 and is not completely eliminated by most enzyme inhibitors. For this reason, any radioimmunoassay using human late pregnancy plasma, extracted or otherwise, should be examined most carefully for the possible action of oxytocinase.

There are two further direct approaches to the problem of measuring maternal oxytocin during human labour: measurement of circulating neurophysin (as in the animal studies already mentioned), and measurement of oxytocin in the urine. The detailed investigations by Legros & Franchimont (1972) and Robinson, Zimmerman & Frantz (1972) showed a progressive rise in the concentration of neurophysin throughout pregnancy reaching maximum levels at term, with no further rise during labour. If this increase reflects oxytocin secretion (at levels, assuming a 1:1 ratio, of up to 4 ng/ml), then the pattern is quite different to that suggested by all previous authors. As to the excretion of oxytocin in urine, recent studies (Boyd & Chard, 1973) have shown that there is no increase (and even perhaps a decrease), during labour. Since an increase in the urinary excretion of oxytocin may be produced by an infusion at a rate of as little as 1 mu. oxytocin/min it would seem that the 'spurt' release during labour contributes little to the total output of oxytocin in the urine.

Indirect evidence

The indirect evidence for the involvement of the maternal posterior pituitary in human labour is less clear-cut than that in animals. For instance, the relative absence of abnormalities of labour in women with diabetes insipidus (Hendricks, 1954), despite clear evidence that oxytocin may be deficient in this condition (Cobo, Bernal & Gaitan, 1972), would suggest that the gland is not of critical importance. Against this, it has been pointed out that the deficiency is relative rather than absolute, and that the functional requirements for oxytocin may be less than those for vaso-pressin (Chau, Fitzpatrick & Jamieson, 1969). Ethanol can depress uterine activity in human labour (Zlatnik & Fuchs, 1972), but it is still not clear whether the effect is on the myometrium or the pituitary. For instance, ethanol has a similar effect on prostaglandin-induced contractions (Karim & Sharma, 1971). Probably the best indirect evidence in man arises from the results of administration of oxytocin in late pregnancy, an experiment performed almost daily in clinical practice. The rate of infusion required to induce labour in the majority of women near term (2–8 mu./min) would not, given the known parameters of metabolic clearance, yield plasma levels in excess of 10 μu./ml. Estimated levels greatly in excess of this in the earlier stages of labour must be viewed with some suspicion. Theobald has proposed that even the low rates noted above are pharmacological rather than physiological, and that the level of endogenous hormone is unlikely to exceed that produced by an infusion at 0·1 mu./min. Further evidence that oxytocin levels are not strikingly raised during labour comes from the elegant experiments of Cobo (1968). By making a continuous recording of intra-mammary pressure, each subject thus acting as her own bioassay, he was unable to detect any oxytocin-like activity during spon-taneous labour, despite the fact that a response could be produced by an injection of as little as 2 mu. oxytocin.

In quantitative terms, our own results obtained with a radioimmunoassay do not disagree radically with the indirect evidence available for man. However, they do differ from the results in animals, including those obtained using the same assay. One major cause of this discrepancy could be the different sampling sites: in animals the jugular system, in women a peripheral vein. Only one study in women, that of Coch et al. (1965) used central venous plasma; the results showed high concentrations comparable to those found in animals, and reaching a peak during the second stage of labour, though the authors expressed reservations as to the exact nature of the material measured. Thus, a major question still remains: do the results obtained from measurements in human peripheral plasma, using a specific radioimmunoassay, reflect a central release of the hormone equivalent to that found in animals, or is there a fundamental difference between species? Further work is required to answer this question.

The cause of oxytocin release

As in animals, the release of oxytocin observed during parturition could be a result of 'Ferguson's reflex'. However, another possibility is raised by recent studies of prostaglandin infusions (Gillespie, Brummer & Chard, 1972), which demonstrate that these may produce a pattern of 'spurt' release virtually identical with that seen in spontaneous labour. Since a similar effect has been demonstrated in male volunteers, it seems likely that prostaglandins have a direct action on the pituitary. There is some evidence that the concentration of circulating prostaglandins may increase during labour (Karim, 1968), and this increase could be an additional mechanism for the activation of the posterior pituitary.

The significance of maternal oxytocin release in woman

The pattern of 'spurt' release of oxytocin reported in human plasma during labour is similar to that found in animals during suckling (McNeilly, 1972) and sexual activity (McNeilly & Ducker, 1972), though there is some doubt as to whether the released oxytocin plays any significant physiological role: for example the denervated transplanted udder of the goat can maintain normal milk yield (Linzell, 1963). However, there is some evidence that intermittent release of oxytocin may have a greater effect on the target organ than a continuous release of the same amount of

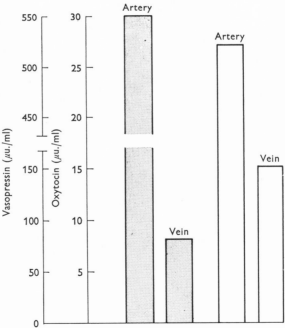

Fig. 3. Oxytocin (white bars) and vasopressin (shaded bars) concentrations in the umbilical artery and vein at the time of delivery in 38 women.

hormone (Caldeyro-Barcia *et al.* 1971); 'spurt' release could, therefore, be biologically effective.

RELEASE OF OXYTOCIN AND VASOPRESSIN BY THE FOETAL POSTERIOR PITUITARY DURING LABOUR

Direct evidence

High concentrations of both oxytocin and vasopressin are consistently found in human umbilical arterial and venous plasma collected at the time of delivery (Chard *et al.* 1970, 1971). The levels are higher in the umbilical

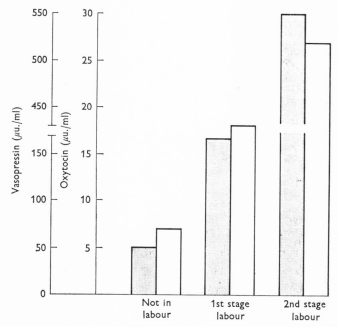

Fig. 4. Oxytocin (white bars) and vasopressin (shaded bars) concentrations in the umbilical arterial plasma of the human foetus at the time of delivery in 51 women. The first stage of labour (four cases) and late pregnancy samples (nine cases) were collected at the time of Caesarean section.

artery than the umbilical vein (Fig. 3) indicating that the origin of the hormone is the foetus itself. The highest concentrations are found at the time of vaginal delivery, the lowest at Caesarean section when the patient was not in labour, and intermediate concentrations at Caesarean section performed during labour. Thus release appears to be associated with the process of labour, being maximal during the expulsive phase (Fig. 4).

The finding of high levels of oxytocin in the umbilical circulation with an arteriovenous difference has been confirmed by P. Kumaresan and his colleagues (personal communication), though it should be noted that these workers also found high concentrations in the maternal circulation. Also in man, Hoppenstein, Miltenberger & Moran (1968) showed high levels of vasopressin in umbilical blood at the time of Caesarean section, and Legros & Franchimont (1972) have found high levels of neurophysin. Seppala, Aho, Tissari & Ruoslahti (1972) have shown that there is an increase in amniotic fluid oxytocin during labour, and that high levels may be found in meconium and in neonatal urine. Our own recent studies have shown high concentrations of oxytocin, vasopressin, and neurophysin in neonatal urine.

In the sheep and the pig, Bell & Robson (1937) demonstrated oxytocin in the foetal pituitary, and suggested that this might play a role in the onset of labour. In the cow, Robinson and coworkers (1972) found that the levels of neurophysin in the foetal circulation were considerably higher than those in the mother at the time of delivery. Furthermore, there was a relative excess of neurophysin I, which is considered to be specific for oxytocin; this suggests that, in contrast to the situation in the human foetus, oxytocin release predominates over vasopressin release. In the foetal sheep and monkey near term, vasopressin secretion is relatively much greater than in the adult (Skowsky, Bashore, Smith & Fisher, 1973) and there is evidence for a very rapid increase in vasopressin secretion in the seven days preceding parturition (Mellor & Slater, 1972). In the guinea-pig, there is a striking increase in the hormone content of the foetal posterior pituitary at or around the time of delivery (Burton & Forsling, 1972). Such an increase does not, of course, necessarily reflect an increased rate of release. Oxytocin can also be found in the circulation of the foetal guinea-pig at the time of delivery, but since there are high levels in maternal blood at this time, the foetal oxytocin could be derived by placental transfer from the mother (Burton, Challis, Illingworth & McNeilly, 1972).

The cause of oxytocin and vasopressin release by the foetus

This is unknown. To provide the basis for the release, maturation of the hypothalamo-neurohypophysial system is obviously essential. Whether this is as clean-cut in relation to parturition in man and other species as it is in the guinea-pig is uncertain. Furthermore, it is not clear whether the release during labour is a primary phenomenon, or is secondary to the effect of uterine contractions. Whether primary or secondary, it is possible to speculate that compression of the foetal head, foetal anoxia, or release of prostaglandins within the foeto-placental unit might play a role as the trigger stimulus. In the sheep foetus, it has been demonstrated that haemorrhage can cause a release of vasopressin and adrenocorticotrophic hormone (ACTH) (Alexander *et al.* 1970).

The significance of oxytocin and vasopressin release by the foetus

Activation of the foetal neurohypophysis during labour and delivery, if it occurs, may simply represent the combination of a mature system and a stimulus provided by the process of labour; the resulting release may not necessarily have functional significance. Yet both oxytocin and vasopressin are highly active biological molecules, and it is difficult to avoid the conclusion that they may exert potent effects. Three sites of action can be suggested: the myometrium, the umbilical circulation, and the foetal anterior pituitary.

In most species, and especially in man, the myometrium at term is exquisitely sensitive to the action of oxytocin. Since oxytocin can readily cross the placenta (Noddle, 1964; P. W. Nathanielsz, unpublished observations), it would appear that the foetal neurohypophysis is well sited to influence uterine activity. Whether oxytocin can act directly when delivered via the umbilical artery, or whether it would require to re-circulate through the mother, is at present unknown.

In common with other blood vessels, the umbilical circulation is sensitive to the vasoconstrictor effect of the posterior pituitary hormones, and in particular to that of oxytocin (Somlyo & Somlyo, 1970). The amounts of oxytocin found in the human umbilical artery at the time of delivery would be enough to cause some 30 per cent of maximal constriction (Altura, Malaviya, Reich & Orkin, 1972). The physiological significance of this effect is not clear.

The anterior pituitary–adrenal system can be stimulated by vasopressin, and there is evidence that vasopressin may be a physiological adrenocorticotrophin releasing factor (Yates *et al.* 1971). In the human foetus, the amount of vasopressin circulating at the time of delivery is pharmacological, and could well be responsible for the release of ACTH and cortisol observed at the same time (Mukherjee & Swyer, 1972). It may be that vasopressin, acting as an adrenocorticotrophin releasing factor may represent the primary afferent link in the activation of the foetal pituitary–adrenal system in those species in which this plays a part in the initiation of labour (Turnbull, 1972). The evidence that hypophysectomy of foetal animals can lead to prolongation of pregnancy (Liggins, Kennedy & Holm, 1967; Chez, Hutchinson, Salazar & Mintz, 1970) could as well implicate the hypothalamus as the anterior pituitary itself. However, this mechanism does not operate in all species, since in some, foetal decapitation has no effect on the timing of delivery (Jost, 1973). Furthermore, the importance of the pituitary–adrenal system varies between species: adrenalectomy of the foetal rhesus monkey has little effect on the length of pregnancy (Mueller-Heubach, Myers & Adamsons, 1972); in the guinea-pig, parturition cannot be induced by administration of corticosteroids (J. R. Challis, personal communication); in women, the increase of corticosteroids in the umbilical circulation at delivery is found in both spontaneous and artificially induced labour (Smith & Shearman, 1972).

CONCLUSIONS

The factors responsible for the initiation and maintenance of labour are multiple, and the importance of any one mechanism is likely to vary between species. Release of active hormone from the posterior pituitary is only one of these factors; the studies discussed here suggest that in man

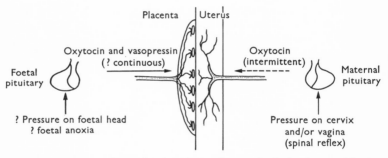

Fig. 5. A tentative scheme for the release of oxytocin during human labour.

there is a contribution from both the mother and the foetus (Fig. 5). The following general conclusions may be drawn: (1) There is little evidence that an increase in circulating maternal oxytocin is responsible for the initiation of labour, though a low fixed level may be an essential permissive factor. However, there is evidence that oxytocin levels increase during labour, especially during the expulsive phase. Maternal oxytocin may, therefore, play a part in the development of uterine contractions during labour, and in the intense activity at the time of delivery. (2) In some species, there appears to be a release of maternal vasopressin during parturition. This may be a response to 'stress', and its functional significance is uncertain. (3) In man and possibly in other species, there is a release of oxytocin from the foetal pituitary during labour. If this has any action it may, like maternal oxytocin, play a part in the maintenance and increase of uterine activity during the process of parturition. This may be of special importance in those species, such as man, in which evidence for other mechanisms is lacking. (4) There is good evidence for foetal vasopressin release during labour in man and for the activation of the foetal hypo-thalamo-neurohypophysial system in late pregnancy in other species. It is conceivable that this may represent part of the primary afferent link for the activation of the foetal pituitary–adrenal system.

ACKNOWLEDGEMENTS

The work reported here was supported by grants from the Spastics Society, the Medical Research Council, and the Board of Governors of St Bartholomew's Hospital. I am most grateful to my coworkers, N. R. H. Boyd, G. L. D. Gibbens and S. A. Hollingsworth.

REFERENCES

ALEXANDER, D. P., BRITTON, H. G., FORSLING, M. L., NIXON, D. A. & RATCLIFFE, J. G. (1970). The release of corticotrophin and vasopressin in the foetal sheep in response to haemorrhage. *J. Physiol., Lond.* **213**, 31–32*P*.

ALLEN, W. E., CHARD, T. & FORSLING, M. L. (1973). Peripheral plasma levels of oxytocin and vasopressin in the mare during parturition. *J. Endocr.* **57**, 175–176.

ALTURA, B. M., MALAVIYA, D., REICH, C. F. & ORKIN, L. R. (1972). Effects of vasoactive agents on isolated human umbilical arteries and veins. *Am. J. Physiol.* **222**, 345–355.

BELL, G. H. & ROBSON, J. M. (1937). The oxytocin content of the foetal pituitary. *Q. Jl exp. Physiol.* **27**, 205–208.

BEYER, C. & MENA, F. (1970). Parturition and lactogenesis in rabbit with high spinal cord transection. *Endocrinology,* **87**, 195–197.

BOYD, N. R. H. & CHARD, T. (1973). Human urine oxytocin levels during pregnancy and labour. *Am. J. Obstet. Gynec.* **115**, 827–829.

BURTON, A. M., CHALLIS, J. R. G., ILLINGWORTH, D. V. & McNEILLY, A. S. (1972). Oxytocin in the plasma of conscious guinea-pigs during parturition. *J. Physiol. Lond.* **226**, 94–95*P*.

BURTON, A. M. & FORSLING, M. L. (1972). Hormone content of the neurohypophysis in foetal, newborn and adult guinea-pigs. *J. Physiol., Lond.* **221**, 6–7*P*.

BURTON, A. M., FORSLING, M. L. & MARTIN, M. J. (1972). Release of neurophysin, oxytocin and vasopressin in the rat. *J. Physiol., Lond.* **217**, 23–24*P*.

CALDEYRO-BARCIA, R., MELANDER, S. & COCH, J. A. (1971). Neurohypophyseal hormones. In *Endocrinology of Pregnancy*, eds. F. Fuchs & A. Klopper, pp. 235–285. New York: Harper & Row.

CHARD, T. (1971). The extraction and concentration of hormones from biological fluids. In *Radioimmunoassay Methods*, eds. K. E. Kirkham & W. M. Hunter, pp. 491–502. London & Edinburgh: Churchill Livingstone.

CHARD, T. (1972). The posterior pituitary in human and animal parturition. *J. Reprod. Fert.* Suppl. **16**, 121–138.

CHARD, T., BOYD, N. R. H., FORSLING, M. L., McNEILLY, A. S. & LANDON, J. (1970). The development of a radioimmunoassay for oxytocin. The extraction of oxytocin from plasma, and its measurement during parturition in human and goat blood. *J. Endocr.* **48**, 223–234.

CHARD, T., HUDSON, C. N., EDWARDS, C. R. W. & BOYD, N. R.H. (1971). Release of oxytocin and vasopressin by the human foetus during labour. *Nature, Lond.* **234**, 352–354.

CHAU, S. S., FITZPATRICK, R. J. & JAMIESON, B. (1969). Diabetes insipidus and parturition. *J. Obstet. Gynaec. Br. Commonw.* **76**, 444–450.

CHEZ, R. A., HUTCHINSON, D. L., SALAZAR, H. & MINTZ, D. H. (1970). Some effects of fetal and maternal hypophysectomy in pregnancy. *Am. J. Obstet. Gynec.* **108**, 643–650.

COBO, E. (1968). Uterine and milk-ejecting activities during human labour. *J. appl. Physiol.* **24**, 317–323.

COBO, E., BERNAL, M. DE & GAITAN, E. (1972). Low oxytocin secretion in diabetes insipidus associated with normal labor. *Am. J. Obstet. Gynec.* **114**, 861–866.

COCH, J. A., BROVETTO, J., CABOT, H. M., FIELITZ, C. A. & CALDEYRO-BARCIA, R. (1965). Oxytocin-equivalent activity in the plasma of women during labour and during the puerperium. *Am. J. Obstet. Gynec.* **91**, 10–17.

FERGUSON, J. K. W. (1941). A study of the motility of the intact uterus at term. *Surgery Gynec. Obstet.* **73**, 359–366.

FISHER, C., MAGOUN, H. W. & RANSON, S. W. (1938). Dystocia in diabetes insipidus. The relation of pituitary oxytocin to parturition. *Am. J. Obstet. Gynec.* **36**, 1–9.

FORSLING, M. L., BOYD, N. R. H. & CHARD, T. (1971). The dissociation of the immunological and the biological activity of oxytocin: in-vivo studies. In *Radioimmunoassay Methods*, eds. K. E. Kirkham & W. M. Hunter, pp. 549–555. London & Edinburgh: Churchill Livingstone.

FUCHS, A.-R. & SAITO, S. (1971). Pituitary oxytocin and vasopressin content of pregnant rats before, during and after parturition. *Endocrinology*, **88**, 574–578.

FUCHS, F. (1971). Endocrinology of labor. In *Endocrinology of Pregnancy*, eds. F. Fuchs & A. Klopper. pp. 306–327. New York: Harper & Row.

GALE, C. C. & McCANN, S. M. (1961). Hypothalamic control of pituitary gonadotrophins. *J. Endocr.* **22**, 107–117.

GIBBENS, D., BOYD, N. R. H. & CHARD, T. (1972). Spurt release of oxytocin during human labour. *J. Endocr.* **53**, liv–lv.

GILLESPIE, A., BRUMMER, H. C. & CHARD, T. (1972). Oxytocin release by infused prostaglandins. *Br. med. J.* i, 543–544.

GLICK, S. M., KUMARESAN, P., KAGAN, A. & WHEELER, M. (1969) Radioimmunoassay of oxytocin. In *Protein and polypeptide Hormones*, ed. M. Margoulies, pp. 81–83. Amsterdam: Excerpta Medica Foundation.

HENDRICKS, C. H. (1954). The neurohypophysis in pregnancy. *Obstetl gynec. Surv.* **9**, 323–341.

HOPPENSTEIN, J. M., MILTENBERGER, F. M. & MORAN, W. H. (1968). The increase in blood levels of vasopressin in infants during birth and surgical procedures. *Surgery Gynec. Obstet.* **127**, 966–974.

JAMES, M. A. R., CHARD, T. & FORSLING, M. L. (1971). The dissociation of biological and immunological activity: in-vitro studies with oxytocin. In *Radioimmunoassay Methods*, eds. K. E. Kirkham & W. M. Hunter, pp. 545–549. London & Edinburgh: Churchill Livingstone.

JOST, A. (1973). Does the foetal hypophyseal–adrenal system participate in delivery in rats and rabbits? In *Foetal and Neonatal Physiology*, eds. R. S. Comline, K. W. Cross, G. S. Dawes & P. W. Nathanielsz, pp. 589–593 London: Cambridge University Press.

KARIM, S. M. M. (1968). Appearance of prostaglandin $F_{2\alpha}$ in human blood during labour. *Br. med. J.* iv, 618–620.

KARIM, S. M. M. & SHARMA, S. D. (1971). The effect of ethyl alcohol on prostaglandins E_1 and $F_{2\alpha}$ induced uterine activity in pregnant women. *J. Obstet. Gynaec. Br. Commonw.* **78**, 251–254.

KUMARESAN, P., ANANDARANGAM, B. & VASICKA, A. (1972). Studies of human oxytocin by radioimmunoassay. In *Abstracts of the IVth International Congress of Endocrinology, Washington, D.C., 1972*, p. 206, Amsterdam: Excerpta Medica Foundation.

LEGROS, J. J. & FRANCHIMONT, P. (1972). Human neurophysin blood levels under normal, experimental and pathological conditions. *Clinical Endocrinology* **1**, 99–113.

LIGGINS, G. C., KENNEDY, P. C. & HOLM, L. W. (1967). Failure of initiation of parturition after electrocoagulation of the pituitary of the fetal lamb. *Am. J. Obstet. Gynec.* **98**, 1080–1086.

LINZELL, J. L. (1963). Some effects of denervating and transplanting mammary glands. *Q. Jl. exp. Physiol.* **48**, 34–60.

McNEILLY, A. S. (1972). The blood levels of oxytocin during suckling and hand-milking in the goat with some observations on the pattern of hormone release. *J. Endocr.* **52**, 177–188.

McNEILLY, A. S. & DUCKER, H. A. (1972). Blood levels of oxytocin in the female goat during coitus and in response to stimuli associated with mating. *J. Endocr.* **54**, 399–406.

McNEILLY, A. S., FORSLING, M. L. & CHARD, T. (1971). Comparison between radioimmunological and biological assay methods and two extraction procedures in the measurement of endogenous oxytocin in plasma. In *Radioimmunoassay Methods*, eds. K. E. Kirkham & W. M. Hunter, pp. 558–560. London & Edinburgh: Churchill Livingstone.

McNEILLY, A. S., MARTIN, M. J., CHARD, T. & HART, I. C. (1972). Simultaneous release of oxytocin and neurophysin during parturition in the goat. *J. Endocr.* **52**, 213–214.

MELLOR, D. J. & SLATER, J. S. (1972). Daily changes in foetal urine and relationships with amniotic and allantoic fluid and maternal plasma during the last 2 months of pregnancy in conscious, unstressed ewes with chronically implanted catheters. *J. Physiol., Lond.* **227**, 503–525.

MUELLER-HEUBACH, E., MYERS, R. E. & ADAMSONS, K. (1972). Effects of adrenalectomy on pregnancy length in the rhesus monkey. *Am. J. Obstet. Gynec.* **112**, 221–226.

MUKHERJEE, J. & SWYER, G. I. M. (1972). Plasma cortisol and adrenocorticotrophic hormone in normal men and non-pregnant women, normal pregnant women and women with pre-eclampsia. *J. Obstet. Gynaec. Br. Commonw.* **79**, 504–512.

NIBBELINK, D. W. (1961). Paraventricular nuclei, neurohypophysis and parturition. *Am. J. Physiol.* **200**, 1229–1232.

NODDLE, B. A. (1964). Transfer of oxytocin from the maternal to the foetal circulation of the ewe. *Nature, Lond.* **203**, 414.

ROBERTS, J. S. & SHARE, L. (1968). Oxytocin in plasma of pregnant, lactating and cycling ewes during vaginal stimulation. *Endocrinology*, **83**, 272–278.

ROBINSON, A. G., ZIMMERMAN, R. A. & FRANTZ, A. G. (1972). Physiologic investigation of posterior pituitary binding proteins neurophysin I and neurophysin II. *Metabolism*, **20**, 1148–1155.

SEPPALA, M., AHO, I., TISSARI, A. & RUOSLAHTI, E. (1972). Radioimmunoassay of oxytocin in amniotic fluid, fetal urine and meconium during late pregnancy and delivery. *Am. J. Obstet Gynec.* **114**, 788–795.

SKOWSKY, W. R., BASHORE, R. A., SMITH, F. G. & FISHER, D. A. (1973). Vasopressin metabolism in the foetus and newborn. In *Foetal and Neonatal Physiology*, eds. R. S. Comline, K. W. Cross, G. S. Dawes & P. W. Nathanielsz, pp. 439–447. London: Cambridge University Press.

SMITH, I. D. & SHEARMAN, R. P. (1972). The relationship of human umbilical arterial and venous plasma levels of corticosteroids to gestational age. *J. Endocr.* **55**, 211–212.

SOMLYO, A. P. & SOMLYO, A. V. (1970). Vascular smooth muscle. II. Pharmacology of normal and hypertensive vessels. *Pharmac. Rev.* **22**, 249–353.

THEOBALD, G. W. (1968). Oxytocin reassessed. *Obstetl gynec. Surv.* **23**, 109–131.

TURNBULL, A. C. (ed.) (1972). *Symposium on Control of Parturition. J. Reprod. Fert.* Suppl. **16**.

YATES, F. E., RUSSELL, S. M., DALLMAN, M. F., HEDGE, G. A., McCANN, S. M. & DHARIWAL, A. P. S. (1971). Potentiation by vasopressin of corticotropin release induced by corticotropin-releasing factor. *Endocrinology*, **88**, 3–15.

ZLATNIK, F. J. & FUCHS, F. (1972). A controlled study of ethanol in threatened premature labor. *Am. J. Obstet. Gynec.* **112**, 610–612.

PROSTAGLANDINS AND HUMAN LABOUR

By ARNOLD GILLESPIE

This review will, as far as possible, be restricted to studies in women in early and advanced pregnancy and I shall draw what I hope are logical conclusions from these observations.

PROSTAGLANDINS IN THE INDUCTION OF LABOUR AND ABORTION

The first report of the use of a prostaglandin (PG) in the induction of labour came from Karim and his coworkers in 1968. Successful induction of labour at or near term by the intravenous infusion of $PGF_{2\alpha}$ in doses varying from 0·025 to 0·05 μg/kg/min were reported (Karim, Trussell, Patel & Hillier, 1968; Karim, Trussell, Hillier & Patel, 1969). In 1970, the first use of PGE_2 in the induction of labour was reported simultaneously by Beazley, Dewhurst & Gillespie (1970) and by Karim et al. (1970). We administered PGE_2 intravenously to 40 women and in 35 vaginal delivery was successfully induced in less than 24 h. The infusion rate varied from 0·5 to 2 μg/min. Many reports of the successful use of prostaglandin $F_{2\alpha}$ and E_2 have followed.

From the clinician's point of view, two questions must be asked: do the prostaglandins have any advantages over oxytocin in the induction of labour? and which prostaglandin is the most effective?

In an attempt to answer these questions, at least in part, a double-blind comparison of PGE_2 and oxytocin (Syntocinon) in the induction of labour was carried out in 1970 at Queen Charlotte's Maternity Hospital, London (Beazley & Gillespie, 1971). Solutions of PGE_2 or oxytocin were administered intravenously in a previously determined randomised pattern to 300 women admitted to the hospital for the induction of labour. The solutions were administered with a slow infusion pump at the rates and doses shown in Table 1. Amniotomy was not carried out before the start of the infusion. Induction of labour was considered to be successful when 6 cm dilatation of the cervix, or delivery, occurred within 12 h of the beginning of the infusion. Eight women were excluded from the trial for technical reasons. The success rate of the two substances in the remaining 292 women was identical. Out of 146 women receiving PGE_2, labour was

[77]

Table 1. *The quantity of prostaglandin E₂ (PGE₂) and oxytocin (Syntocinon) delivered at each speed of the constant infusion pump. (Reproduced with permission from Beazley and Gillespie, 1971.)*

	Delivery per minute	
Pump setting (inch/min)	Oxytocin (50 mu./ml) (mu.)	PGE₂ (5 μg/ml) (μg)
1/320	2·1	0·21
1/160	4·2	0·42
1/80	8·4	0·84
1/40	16·8	1·67
1/20	33·5	3·55
1/10	67·0	6·70

induced successfully in 106; out of 146 women receiving Syntocinon, labour was induced successfully in 107.

Having established that PGE_2 was as effective as oxytocin in the induction of labour near term, we compared the relative efficiencies of PGE_2 and $PGF_{2\alpha}$. Restricted supplies of $PGF_{2\alpha}$ at the time limited the trial to 30 women. The compounds were administered in an incremental dosage

Table 2. *The rate of infusion of prostaglandin E₂ (PGE₂) and F₂ₐ (PGF₂ₐ) and the time for which each dose was infused to induce labour*

Infusion rate (μg/min)		Time of infusion at each rate (h)
PGE₂	PGF₂ₐ	
0·3125	2·5	½
0·625	5·0	½
1·25	10·0	1
2·50	20·0	4
5·00	40·0	4

regime (Table 2). No difference in efficiency was apparent between the two substances but at these rates of administration $PGF_{2\alpha}$ caused gastrointestinal side effects in a significant number of women.

Table 3 shows the results of a comparison of PGE_2, $PGF_{2\alpha}$ and oxytocin carried out by Karim (1971). From these results it would appear that PGE_2 is far more efficient than oxytocin. It should, however, be noted

Table 3. *Result of double-blind clinical trial with prostaglandin E₂ (PGE₂) and F₂ₐ (PGF₂ₐ) and oxytocin for induction of labour. (Reproduced with permission from Karim, 1971.)*

Substance	Dose range	Number of patients	Successful inductions
PGE₂	0·3–1·2 μg/min	100	96
PGF₂ₐ	2·5–10 μg/min	100	67
Oxytocin	2–8 mu./min	100	56

that the maximum infusion rate for oxytocin was 8 mu./min which in many centres would be considered an inadequate dose. More recent reports (Anderson, Hobbins & Speroff, 1972; Vakhariya & Sherman, 1972) confirm our findings that PGE_2, $PGF_{2\alpha}$ and oxytocin are equally effective in inducing labour when adequate doses of each substance are used.

The relationship between dose and maternal weight, parity and length of gestation

The doses of PGE_2 which induced labour successfully were examined in relation to maternal weight, parity and length of gestation and no correlation was found between dose and any of these parameters (Gillespie, 1972). This is surprising in respect of maternal weight. After intravenous injection of prostaglandin, a large percentage is inactivated by the lungs at the first circulation. It can be assumed that only the PG that is not inactivated acts on the uterus. It seemed probable that the inactivating mechanism would become saturated during continuous intravenous infusion; subsequent work measuring the concentration of prostaglandin in the blood during continuous infusion supports this assumption. It was also surprising that the effective dose of PG was not related to parity or length of gestation. It is usually assumed that labour is easier to induce in a multigravida than a primigravida and that it is easier the nearer the period of gestation is to term. Our results did not support these assumptions. If one can draw any conclusion from the lack of correlation between the dose of prostaglandin and maternal weight, parity and period of gestation, then perhaps it is that other factors exist which are more important than the prostaglandins in determining the degree of uterine activity.

Other routes of administration

Oral administration

Recently it has been shown that the prostaglandins are effective uterine stimulants when administered orally. The results of Karim's original work are shown in Table 4. Of 80 women receiving PGE_2, 79 had labour induced successfully. When $PGF_{2\alpha}$ was administered orally to 20 women, 16 had labour induced successfully (Karim & Sharma, 1971 a).

The effective dose of PGE_2 varied from 0·5 to 1·5 mg and that of $PGF_{2\alpha}$ from 5 to 15 mg. The dose was repeated at 2-hourly intervals since this was the average time for which one dose was effective. Karim has recently reported the results of investigations of 1000 women in whom labour has been induced or accelerated by the oral administration of PGE_2. About

Table 4. *Method of delivery in 80 women in whom labour was induced with oral prostaglandin E_2. (Reproduced with permission from Karim and Sharma, 1971a.)*

Method of delivery	Primigravida	Multigravida
Spontaneous vaginal	25	44
Operative vaginal	3	5
Caesarean section	2*	1†

* Cephalopelvic disproportion.
† Failed induction after 48 h.

10 per cent of the women experienced diarrhoea and/or vomiting, but no serious side effects were reported (S. M. M. Karim, unpublished results, Brook Lodge Symposium on Prostaglandins, June 1972, Upjohn Company).

Wallace Barr has also used prostaglandins orally to induce labour (Barr & Naismith, 1972; Brook Lodge Symposium on Prostaglandins, June 1972, Upjohn Company). Defining successful induction of labour as an induction in under 18 h, the following results were obtained: out of 50 women receiving PGE_2, induction of labour was successful in 32; out of 50 women receiving $PGF_{2\alpha}$, induction was successful in 33. Most of the women receiving $PGF_{2\alpha}$ had diarrhoea and nine vomited.

The place of orally administered prostaglandins in the induction of labour is not yet clear. Although the success rate is high, and even if the gastro-intestinal side effects can be reduced, I cannot believe that this method will be popular with obstetricians who have come to rely on the precise control of uterine activity afforded by intravenously administered oxytocic agents.

Intravaginal administration

Labour and abortion can be induced by introducing prostaglandin, contained in a suitable vehicle, into the posterior fornix of the vagina.

Table 5. *Induction of abortion by the vaginal administration of prostaglandin E_2 (PGE_2) or $F_{2\alpha}$ ($PGF_{2\alpha}$) to 45 women (Reproduced with permission from Karim and Sharma, 1971b.)*

Gravida	Prostaglandin	Gestation (weeks)	Number of women	Average induction– abortion interval (h)
1	PGE_2	9–13	4	15
1		14–20	11	16
2–7		7–13	2	11
2–9		14–23	13	9
1	$PGF_{2\alpha}$	9	1	$28\frac{1}{2}$
1		14–22	5	18
2–6		10	1	12
2–8		14–21	8	$11\frac{1}{2}$

Typical results are shown in Table 5 (Karim & Sharma, 1971 b). Side effects occur which are similar to those seen when the prostaglandin is administered intravenously. Kirton, using rhesus monkeys, has shown (by excluding the uterus from contact with the prostaglandin in the vaginal fornix by covering the cervix with collodion) that the uterine response results from absorption of the prostaglandin from the vagina into the systemic circulation and not by a direct stimulatory effect of the PG on the uterus by way of the cervical canal (Kirton, 1972).

Intra-amniotic injection

The intra-amniotic injection of prostaglandins has no place in the induction of labour, but it is an effective means of terminating mid-trimester pregnancy. A dose of $PGF_{2\alpha}$ of the order of 25 mg is used and success rates of about 90 per cent are obtained. Many women require a second injection 24 h after the first (Bygdeman, Toppozada & Wiqvist, 1971; Karim & Sharma, 1971 c; Toppozada, Bygdeman & Wiqvist, 1971).

Blood concentrations during administration of prostaglandin

It has been established that the prostaglandins can increase the activity of the uterus when administered by a variety of routes. It now remains for investigators to establish the place of these routes of administration in a given clinical situation.

Since general side-effects of greater or lesser degree are observed regardless of the route of administration, systemic absorption of the prostaglandin must always occur. Using radioimmunoassays, concentrations of circulating prostaglandins during intravenous infusions of PG at different rates have been determined. Caldwell, Anderson, Hobbins & Speroff (1972) studied four women receiving an intravenous infusion of $PGF_{2\alpha}$ at a rate of 25 μg/min. The concentration of PGF in the blood rose within 10 min and reached 2–3 ng/ml at 30 min. When the infusion rate was doubled to 50 μg/min, only a slight rise in prostaglandin concentration occurred and a plateau was reached 1–2 h after the beginning of the infusion. Investigators at the Karolinska Institute have demonstrated similar findings: $PGF_{2\alpha}$ concentration rose rapidly to a maximum and then remained steady. A clear correlation was seen between the level of circulating $PGF_{2\alpha}$ and the incidence of diarrhoea and vomiting (Green *et al.* 1972). On cessation of the infusion, a rapid drop to baseline concentrations of circulating $PGF_{2\alpha}$ occurred within 5–10 min.

The concentrations of circulating PGF in women being aborted by the intravaginal application of 50 mg $PGF_{2\alpha}$ have also been studied by groups at Yale and at the Karolinska Institute. The Yale group found that three women who aborted successfully developed higher blood concentrations (of the order of 4 ng PGF/ml) than two women who did not (1–2 ng/ml). In general, the incidence of side-effects corresponded with the period when peak concentrations of PGF were observed. The Karolinska group found that the highest circulating concentration of $PGF_{2\alpha}$ (3·3 ng/ml) occurred 2 h after administration of the pessary. The concentration then progressively declined to basal levels during the subsequent 8 h. Uterine tonus and incidence of side effects were maximal 1·5 to 3 h after administration of the PG, i.e. when the concentration of $PGF_{2\alpha}$ in the blood was highest.

Using an intra-amniotic injection of 10 mg $PGF_{2\alpha}$ followed by a further injection of 10 mg several hours later, the Yale group was unable to demonstrate a rise in the concentration of PGF in blood samples taken throughout a 24 h period. No side effects were observed and abortion was not achieved. With an increased dose (40 mg $PGF_{2\alpha}$) in a single injection, 30 women were successfully aborted. Low concentrations (1–2 ng/ml) were recorded in the 4 h after the injection; concentrations then rose to a peak of about 5 ng/ml at 6–8 h. Concentrations of PG in all the patients had fallen to baseline levels 24 h after the injection. These results suggest a very slow release from the amniotic cavity; $PGF_{2\alpha}$ can be measured in mg quantities in the amniotic fluid 8–12 h after the initial injection. It is possible that the PGF being measured in the blood is endogenous prostaglandin and not prostaglandin which has been passively transferred from the amniotic cavity.

In summary, the measurements of circulating PGF during administration of $PGF_{2\alpha}$ by the oral, intravaginal or intra-amniotic routes have confirmed that to be effective, the prostaglandin must be absorbed in large enough amounts to produce a concentration in the blood similar to that seen when the compound is administered intravenously. Before administration of $PGF_{2\alpha}$, concentrations of endogenous prostaglandin in the blood are always less than 1 ng/ml. Effective uterine contractions occur at different concentrations, but usually when the circulating $PGF_{2\alpha}$ concentrations are about 3–5 ng/ml. The incidence of side effects corresponds with the peak concentrations of circulating $PGF_{2\alpha}$.

Although several groups of investigators are working on the problem, very little has been published about the levels of circulating PG during spontaneous labour in women. In the non-pregnant state, most women have concentrations of $PGF_{2\alpha}$ in the blood ranging from 0·25–0·5 ng/ml, although in some women it is undetectable (Cernosek, Morrill & Levine, 1972).

In 1968, Karim, using a biological assay method, estimated the levels of $PGF_{2\alpha}$ in the peripheral blood of women in labour and related these levels to the occurrence of uterine contractions. His results are summarised in Fig. 1 (Karim, 1968). Initial work in our laboratory tends not to support these correlations. In this context it is noteworthy that the concentrations of PG reached during continuous infusion were approximately 5 ng/ml and effective uterine contractions in mid-trimester occurred at this level. It can be expected that lower blood concentrations would effectively

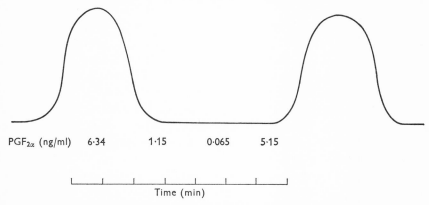

PGF$_{2\alpha}$ (ng/ml)   6·34   1·15   0·065   5·15

Time (min)

Fig. 1. Concentrations of prostaglandin $F_{2\alpha}$ ($PGF_{2\alpha}$) in the peripheral blood of women during labour relative to uterine contractions. (After Karim, 1968.)

stimulate the uterus near term and Karim's figures would therefore seem unnecessarily high. Indeed, marked gastro-intestinal side effects would be expected when blood levels of $PGF_{2\alpha}$ are in the region of 6 ng/ml. Such side effects are not seen in normal labour. It is possible that the measurement of prostaglandin concentrations in peripheral blood will be unrewarding since the prostaglandins probably act locally in the uterus and any detected in peripheral blood would thus represent a 'spill-over'.

ORIGIN OF THE PROSTAGLANDINS

The prostaglandins have been found in almost every human tissue so far examined. The decidua is a particularly rich source and levels of several hundred ng/g tissue have been reported (Karim & Devlin, 1967). Much lower concentrations have been found in liquor amnii. It has been suggested that the concentrations of PG in the liquor amnii of women in labour are higher than those in the liquor amnii of women not in labour (Karim & Devlin, 1967). This could be a secondary phenomenon resulting from an increase in PG production by the decidua at the time of labour. The stimulus for this increased production is not known; indeed, it must be admitted that the evidence for its occurrence is not strong. It has been shown that distension of the uterus of the guinea-pig causes a release of PG (Horton, Thompson, Jones & Poyser, 1971). On the other hand, it is possible that oestrogen could be the controlling stimulus.

PROSTAGLANDINS AND PROGESTERONE

It has been established that in some farm animals and in rodents, $PGF_{2\alpha}$ can have a luteolytic effect. This has not been shown to be true in women. The effect of PG on the production of placental progesterone has not been elucidated. Most workers have shown no change in plasma progesterone levels when prostaglandin is administered to women to elicit labour or abortion. A fall in plasma progesterone concentration occurs after the products of conception have been expelled from the uterus (Bygdeman & Wiqvist, 1971; Speroff, 1971). Wiest & Csapo (1972), however, have demonstrated a decrease in plasma progesterone in women who aborted after administration of $PGF_{2\alpha}$ intravenously or intra-amniotically.

The concentrations of progesterone in the peripheral blood may not, however, reflect what is occurring locally in the myometrium.

PROSTAGLANDINS AND OXYTOCIN

Endogenous oxytocin

The levels of endogenous plasma oxytocin during the intravenous infusion of PGE_2 or $PGF_{2\alpha}$ were studied recently and I am indebted to Dr Tim Chard, who carried out the oxytocin assays using the sensitive radio-immunoassay method developed by himself and his colleagues (Chard et al. 1970). The initial study was carried out on seven women who had been admitted to hospital for the induction of labour. Blood samples (10 ml) were taken immediately before PG infusion, $\frac{1}{2}$ to 1 h later, and then 4 h and 8 h after the start of the infusion. In a further study, blood samples were taken from 15 women before infusion of the PG and hourly for up to 8 h or until delivery occurred.

Urine was collected during the infusion period. The total volume was measured, 5 ml samples were taken and acidified with 1 ml M-HCl, and the sample stored at $-15\ ^{\circ}C$. Assays for oxytocin were carried out as described by Boyd & Chard (1971).

The results are shown diagrammatically in Fig. 2 but may be summarised as follows: (i) Oxytocin was detected in the plasma of 19 out of 22 women receiving a PG infusion; 139 plasma samples were collected and in 60 (43 per cent) oxytocin was detected. (ii) Oxytocin release was intermittent.

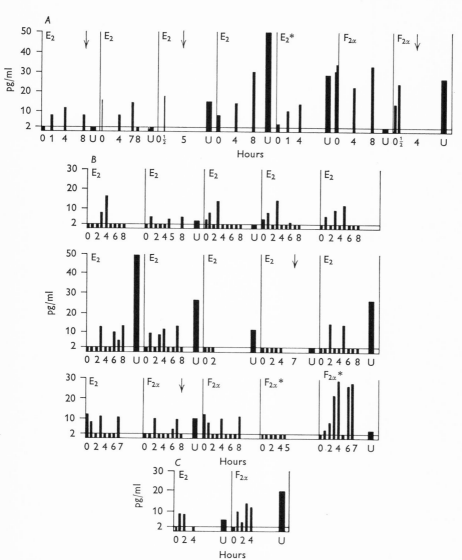

Fig. 2. Plasma and urinary oxytocin concentrations during prostaglandin (PG) infusion. Each section represents one patient. A, seven women, blood samples were taken approximately every 4 h; B, 15 women, blood samples taken hourly; C, two adult male volunteers receiving PG, i.v. for 4 h. Areas below 2 pg/ml (limit of sensitivity) indicate blood samples taken but oxytocin not detected. Delivery is indicated by an arrow; U = urine; * = intrauterine death. (Reproduced with permission from Gillespie, A., Brummer, H. C. & Chard, T. (1972). Oxytocin release by infused prostaglandin. *Br. med. J.* i, 543–544.)

(iii) Oxytocin was detected in the plasma of two of the three women in whom intrauterine death of the foetus had occurred. (iv) Oxytocin was not detected in the plasma of one woman in whom the PG infusion failed to produce uterine contractions.

The intermittent nature of the oxytocin release found in this series of observations corresponds closely with that found in spontaneous first stage labour (Gibbens, Boyd & Chard, 1972).

Chard *et al.* (1970) have demonstrated a higher concentration of oxytocin in the foetal than in the maternal circulation during spontaneous labour and suggest that the foetus plays an important part in the initiation of normal labour by the release of oxytocin from its pituitary gland. It is possible that in the group of women receiving PG described above, the infused prostaglandin stimulated the release of oxytocin from the foetal pituitary gland. The finding of oxytocin in the maternal plasma even when the foetus was dead suggests, however, that in these cases at least, the oxytocin found in the plasma during PG infusion was of maternal origin.

In 1941, Ferguson demonstrated that in the rabbit, dilatation of one cervix enhanced the contractions of the contralateral uterus. Injection of oxytocin or electrical stimulation of the pituitary stalk produced similar results and spinal section or cauterisation of the pituitary stalk abolished the reflex response. To eliminate the possibility that this reflex caused the appearance of endogenous oxytocin in the plasma merely as a response, secondary to cervical stretching by the foetal presenting part resulting from the uterine contractions stimulated by the prostaglandin, an infusion of prostaglandin was administered to two men. Blood and urine samples were collected as before. Oxytocin was not detected in the plasma of either subject before the PG infusion, but after the infusion, oxytocin was found in concentrations similar to those seen in the plasma of women during prostaglandin-induced labour (Fig. 2).

Very similar results have been obtained in a preliminary examination of four women in the mid-trimester of pregnancy. None of the women had detectable oxytocin in plasma samples taken before PG infusion, but in all the women oxytocin was present in the majority of the plasma samples collected hourly for 8 h after the beginning of the infusion. The release was intermittent as before and in the range seen in late pregnancy.

Thus, it seems that intravenous infusion of prostaglandin brings about the release of oxytocin from the maternal pituitary gland. In this connection, it is of interest to note that uterine contractions induced by an infusion of PG can be inhibited by intravenous ethanol (Karim & Sharma, 1971 *d*). Ethanol is known to inhibit the release of posterior pituitary hormones (Fuchs, 1971).

Exogenous oxytocin

. *Enhancement*

It has been demonstrated that after exposure to the ketonic E prostaglandins, guinea-pig myometrium *in vitro* shows an increased response to agonists such as oxytocin and $PGF_{2\alpha}$ (Clegg, Hall & Pickles, 1966; Pickles, Hall, Clegg & Sullivan, 1966). This response is termed 'enhancement'. The enhanced response can be demonstrated for about 1 h after the exposure of the tissue to the PGE. The response can occur in depolarising media, suggesting that the prostaglandin produced the enhancement by an action at an intracellular site, possibly by facilitation of intracellular excitation–contraction coupling, rather than facilitating cell-to-cell conduction. The F prostaglandins do not elicit the phenomenon of enhancement.

Enhancement and potentiation – experiments in vivo

Experiments were designed to see if 'enhancement' could be demonstrated *in vivo*. Seven women who were admitted to the hospital for termination of mid-trimester pregnancies by hysterotomy were offered the possibility of termination by the vaginal route. Their informed consent to the procedures to be carried out was obtained.

An intravenous infusion of 0·9 per cent sterile sodium chloride solution was established using a constant infusion pump. A fluid-filled polyethylene catheter (Portex 90 cm, 100/380/150) was introduced transabdominally, under local anaesthesia, into the amniotic cavity. The catheter was connected to a pen recorder via a semiconductor strain gauge pressure transducer and appropriate amplifier.

It was established that 'enhancement' did occur in the intact pregnant uterus. When the infusion of PGE_2 was administered for $\frac{1}{2}$ to 1 h and then stopped, a greater response to a given dose of oxytocin was observed after the PGE_2 infusion than before. The enhanced response could be obtained for 60–90 min after the end of the PGE_2 infusion.

The essence of the 'enhancement' phenomenon is the serial exposure of the myometrium to the compounds. Prostaglandin E_2 initiates some change in the myometrial cell which persists for over an hour after administration of PGE_2 has ceased. This change renders the cell more responsive to oxytocin.

Oxytocin and prostaglandins administered simultaneously gave the more usual response of 'potentiation'. By studying various combinations of doses, a suitable 'working combination' of oxytocin and prostaglandin was arrived at. Figure 3, an intra-amniotic pressure tracing from a woman 16 weeks pregnant, shows both 'enhancement' and 'potentiation'. Traces *A*, *B* and *C* show the response of the uterus to an infusion of oxytocin progressively increasing from 16 mu. to 128 mu./min. The rapid fall off in response when the oxytocin infusion ceased is seen towards the end of trace *C*. In trace *D* there is a progressively increasing response to the infusion of PGE_2 at rates of 2·5 μg/min and at 5 μg/min. In trace *E* at

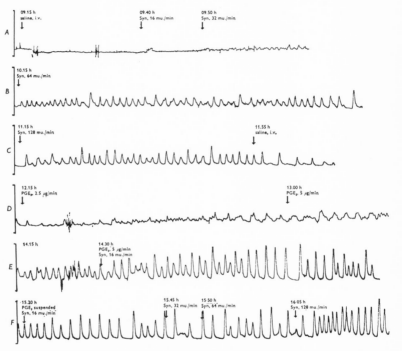

Fig. 3. Intra-amniotic pressure tracing from a woman 16 weeks pregnant showing both 'enhancement' and 'potentiation' in response to infusions of oxytocin (Syntocinon) and prostaglandin E_2 (PGE_2) administered separately and together. Full-scale vertical deflection 100 mmHg.

14.30 h 16 mu. oxytocin/min were added to the infusion of PGE_2 (5 μg/min). The sharply increased amplitude of the uterine contractions (greater than the sum of the amplitudes of contractions achieved with PGE_2 (5 μg/min) or oxytocin (16 mu./min) separately) can be seen in the remainder of trace *E*. These results illustrate the phenomenon of 'potentiation'.

At the beginning of trace *F* the PGE infusion was stopped. The response to 16–128 mu. oxytocin/min was then greater than that seen in traces *A*, *B* and *C* before prostaglandin infusion. Note that the enhancement still persisted an hour after the end of the prostaglandin infusion. By 90 min the response was decreasing but could be re-established by another short infusion of PGE_2. Only one patient was aborted by this technique alone. In practice, the necessity to repeat the infusion of PGE_2 every 1 to $1\frac{1}{2}$ h proved too tedious to be applicable clinically.

Prostaglandin $F_{2\alpha}$ does not enhance the response to subsequently infused oxytocin *in vitro*. No attempt has been made to confirm this observation *in vivo*.

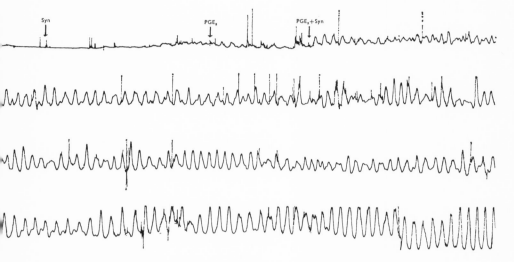

Fig. 4. Intra-amniotic pressure tracing from a woman 16 weeks pregnant during an infusion of 0·8 μg prostaglandin E_2 (PGE_2)/min and 64 mu. oxytocin (Syntocinon)/min separately and together. Full-scale vertical deflection 100 mmHg. Each trace covered a period of $1\frac{1}{2}$ h. The recording was 'offset' by 30 mmHg in the final 15 min of trace D to show the 100 mmHg amplitude contractions occurring at that time.

The activity of the uterus of a woman 16 weeks pregnant is recorded in Fig. 4. Traces A, B, C and D are continuous and each lasted $1\frac{1}{2}$ h. Basal uterine activity is shown in the first 10 min of trace A. A slight increase in activity occurred during the 20 min infusion of oxytocin at 64 mu./min and during the 20 min infusion of 0·8 μg PGE_2/min that followed. In the last third of trace A the potentiated myometrial response to the simultaneous infusion of the two compounds can be seen. Traces B, C and D show the progessive evolution of uterine activity observed when this combination of compounds is given. The amplitude of contractions and the basal uterine tone can be seen to increase. The recording was 'offset' by 30 mmHg in the final 15 min of trace D to show the 100 mmHg amplitude contractions occurring at this time. The contractions were recorded for a further ½ h until the membranes ruptured. Complete abortion occurred 3½ h later.

The clinical usefulness of the potentiation by the prostaglandins of uterine response to oxytocin has recently been tested. Women presenting for termination of pregnancy in the mid-trimester were interviewed, the procedure was explained and if they agreed to participate in the trial, they were randomly allocated to receive either PGE_2 at the rate of 1 μg/min or $PGF_{2\alpha}$ at the rate of 10 μg/min, or either of these prostaglandins plus 64 mu. oxytocin/min. This amount of prostaglandin is one fifth the usual abortifacient dose.

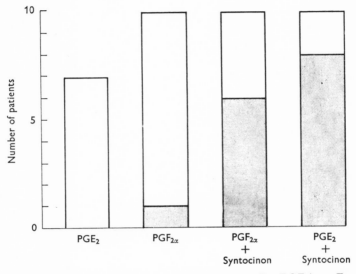

Fig. 5. Abortion rate in women receiving prostaglandin E_2 (PGE_2) or $F_{2\alpha}$ ($PGF_{2\alpha}$) separately or with oxytocin (Syntocinon). The stippled area on the histogram represents the number of women in whom abortion was successfully induced; the unstippled areas, women who failed to abort.

The results are shown in Fig. 5. It is clear that a prostaglandin infusion alone was an ineffective abortifacient at the doses used but that the addition of oxytocin allowed the synergistic response of the uterus to develop to produce an efficient technique for the induction of mid-trimester abortion without systemic side effects.

CONCLUSION

(1) Prostaglandins of the E and F series stimulate contractions of the human pregnant uterus at all stages of gestation and can be used clinically

to induce labour and abortion when administered by a variety of routes.

(2) Whatever the route of administration, systemic absorption takes place and generalised side effects will occur, their degree roughly paralleling the increase in the blood concentration of the prostaglandin.

(3) The decidua is rich in prostaglandins.

(4) Prostaglandins effect the release of oxytocin from the maternal pituitary.

(5) Human pregnant myometrium shows the phenomena of 'enhancement' and 'potentiation' of response when exposed to prostaglandins and oxytocin.

Accepting these premises, the following hypothesis could be proposed: in response to the appropriate stimulus, a release of prostaglandins from the decidua heralds the onset of labour and causes a release of oxytocin from the maternal pituitary, the effect of which is, in turn, enhanced in the myometrium by the local action of the decidual prostaglandins resulting in the evolution of uterine activity that we call normal labour.

REFERENCES

ANDERSON, G., CORDERO, L., HOBBINS, J. & SPEROFF, L. (1971). Clinical use of prostaglandins as oxytocin substances. *Ann. N.Y. Acad. Sci.* **180**, 499–512.

ANDERSON, G. G., HOBBINS, J. C. & SPEROFF, L. (1972). Intravenous prostaglandins E_2 and $F_{2\alpha}$ for the induction of term labour. *Am. J. Obstet. Gynec.* **112**, 382–386.

BARR, W. & NAISMITH, W. C. M. K. (1972). Oral prostaglandins in the induction of labour. *Br. med. J.* ii, 188–191.

BEAZLEY, J. M., DEWHURST, C. J. & GILLESPIE, A. (1970). Induction of labour with prostaglandin E_2. *J. Obstet. Gynaec. Br. Commonw.* **77**, 193–199.

BEAZLEY, J. M. & GILLESPIE, A. (1971). Double blind trial of prostaglandin E_2 and oxytocin in induction of labour. *Lancet*, i, 152–155.

BOYD, N. R. H. & CHARD, T. (1971). A method for extracting oxytocin from urine. In *Radioimmunoassay Methods*, eds. K. E. Kirkham & W. M. Hunter, pp. 512–517. London & Edinburgh: Churchill Livingstone.

BYGDEMAN, M., TOPPOZADA, M. & WIQVIST, N. (1971). Induction of mid-trimester abortion by intra-amniotic administration of prostaglandin F_{2x}. *Acta physiol. scand.* **81**, 415–416.

BYGDEMAN, M. & WIQVIST, N. (1971). Early abortion in the human. *Ann. N.Y. Acad. Sci.* **180**, 473–479.

CALDWELL, B. V., ANDERSON, G. G., HOBBINS, J. C. & SPEROFF, L. (1972). F prostaglandin levels in women receiving $PGF_{2\alpha}$ for therapeutic abortion. In *W.H.O. 3rd Conference on Prostaglandins in Fertility Control*, eds. S. Bergström, K. Green & B. Samuelsson, pp. 182–187. Karolinska Institutet, Stockholm.

CERNOSEK, R. M. G., MORRILL, L. M. & LEVINE, L. (1972). Prostaglandin F_{2x} levels in peripheral sera of man. *Prostaglandins*, **1**, 71–80.

CHARD, T., BOYD, N. R. H., FORSLING, M. L., McNEILLY, A. S. & LANDON, J. (1970). The development of a radioimmunoassay for oxytocin: the extraction of oxytocin from plasma, and its measurement during parturition in human and goat blood. *J. Endocr.* **48**, 223–234.

CLEGG, P. C., HALL, W. J. & PICKLES, V. R. (1966). The action of ketonic pro-staglandins on the guinea pig myometrium. *J. Physiol., Lond.* **183**, 123–144.

FERGUSON, J. K. W. (1941). Study of the motility of the intact uterus at term. *Surgery Gynec. Obstet.* **73**, 359.

FUCHS, F. (1971). Endocrinology of labor. In *Endocrinology of Pregnancy*, eds., F. Fuchs & A. Klopper, pp. 306–327. New York: Harper & Row.

GIBBENS, D., BOYD, N. R. H. & CHARD, T. (1972). Spurt release of oxytocin during human labour. *J. Endocr.* **53**, liv–lv.

GILLESPIE, A. (1972). Factors affecting the dose of prostaglandin E2 and Syntocinon required to induce labour. *J. Obstet. Gynaec. Br. Commonw.* **79**, 135–138.

GREEN, F., BEGUM, F., BYGDEMAN, M., TOPPOZADA, M. & WIQVIST, N. (1972). Analysis of prostaglandin F2α and metabolites following intravenous intra-amniotic and vaginal administration of prostaglandin F2α. In *W.H.O. 3rd Conference on Prostaglandins in Fertility Control*, eds. S. Bergström, K. Green & B. Samuelsson, pp. 189–200. Karolinska Institutet, Stockholm.

HORTON, E. W., THOMPSON, C., JONES, R. & POYSER, N. (1971). Release of prostaglandins. *Ann. N.Y. Acad. Sci.* **180**, 351–362.

KARIM, S. M. M. (1968). Appearance of prostaglandin F2α in human blood during labour. *Br. med. J.* iv, 618–620.

KARIM, S. M. M. (1971). Action of prostaglandin in the pregnant woman. *Ann. N.Y. Acad. Sci.* **180**, 483–498.

KARIM, S. M. M. & DEVLIN, J. (1967). Prostaglandin content of amniotic fluid during pregnancy and labour. *J. Obstet. Gynaec. Br. Commonw.* **74**, 230–234.

KARIM, S. M. M., HILLIER, K., TRUSSELL, R. R., PATEL, R. C. & TAMUSANGE, S. (1970). Induction of labour with prostaglandin E2. *J. Obstet. Gynaec. Br. Commonw.* **77**, 200–210.

KARIM, S. M. M. & SHARMA, S. D. (1971a). Oral administration of prostaglandins for the induction of labour. *Br. med. J.* i, 260–262.

KARIM, S. M. M. & SHARMA, S. D. (1971b). Therapeutic abortion and induction of labour by the intravaginal administration of prostaglandins E2 and F2α. *J. Obstet. Gynaec. Br. Commonw.* **78**, 294–300.

KARIM, S. M. M. & SHARMA, S. D. (1971c). Second trimester abortion with single intra-amniotic injection of prostaglandin E2 or F2α. *Lancet*, ii, 47.

KARIM, S. M. M. & SHARMA, S. D. (1971d). The effect of ethyl alcohol on prosta-glandins E2- and F2α-induced uterine activity in pregnant women. *J. Obstet. Gynaec. Br. Commonw.* **78**, 251–254.

KARIM, S. M. M., TRUSSELL, R. R., HILLIER, K. & PATEL, R. C. (1969). Induction of labour with prostaglandin F2α. *J. Obstet. Gynaec. Br. Commonw.* **76**, 769–782.

KARIM, S. M. M., TRUSSELL, R. R., PATEL, R. C. & HILLIER, K. (1968). Response of pregnant human uterus to prostaglandin F2α induction of labour. *Br. med. J.* iv, 621–623.

KIRTON, K. T. (1972). The role of prostaglandins in reproduction in sub-human primates. In *W.H.O. 3rd Conference on Prostaglandins in Fertility Control*, eds. S. Bergström, K. Green & B. Samuelsson, pp. 208–216. Karolinska Institutet, Stockholm.

PICKLES, V. R., HALL, W. J., CLEGG, P. C. & SULLIVAN, T. J. (1966). Some experi-ments on the mechanisms of action of prostaglandins on the guinea-pig and rat myometrium. *Mem. Soc. Endocr.* **14**, 89–103.

SPEROFF, L. (1971). Discussion of session on prostaglandins in female reproductive physiology. *Ann. N.Y. Acad. Sci.* **180**, 513–517.

TOPPOZADA, M., BYGDEMAN, M. & WIQVIST, N. (1971). Induction of abortion by intra-amniotic administration of prostaglandin F2α. *Contraception*, **4**, 293–332.

VAKHARIYA, V. R. & SHERMAN, A. I. (1972). Prostaglandin $F_{2\alpha}$ for induction of labor. *Am. J. Obstet. Gynec.* **113**, 212–220.

WIEST, W. G. & CSAPO, A. I. (1972). On the correlation between prostaglandin $F_{2\alpha}$-induced progesterone withdrawal and the evolution of pressure and abortion. Brook Lodge Symposium on the Prostaglandins, June 12–14. Upjohn Company.

SPONTANEOUS OR DEXAMETHASONE-INDUCED PARTURITION IN THE SHEEP AND GOAT: CHANGES IN PLASMA CONCENTRATIONS OF MATERNAL PROSTAGLANDIN F AND FOETAL OESTROGEN SULPHATE

By W. B. CURRIE, M. S. F. WONG, R. I. COX
AND G. D. THORBURN

INTRODUCTION

The elegant demonstration by Liggins, Kennedy & Holm (1967) of the involvement of the foetus in the initiation of parturition has led to a rapid accumulation of information on the endocrine events associated with parturition in the sheep. Following Liggins' (1968) suggestion that activation of the pituitary–adrenal system of the foetus is responsible for the initiation of parturition, Bassett & Thorburn (1969) have shown that plasma corticosteroids in the ovine foetus increase dramatically to reach maximal concentrations on the day of parturition and this has been verified by Comline, Nathanielsz, Paisey & Silver (1970). Evidence presented by Thorburn, Nicol, Bassett, Shutt & Cox (1972) and Bassett & Thorburn (1972) suggests that the increase in foetal corticosteroid secretion is responsible for the drop in maternal progesterone concentrations observed in both sheep and goats before parturition. During late pregnancy in the ewe, progesterone is largely secreted by the placenta and the rapid decrease in plasma concentration that generally occurs before parturition (Bassett, Oxborrow, Smith & Thorburn, 1969; Fylling, 1970; Bassett & Thorburn, 1972) probably reflects an action of the foetal corticosteroids on progesterone synthesis or secretion by the placenta (see discussion of other possibilities by Liggins, 1969). In the goat, on the other hand, the corpus luteum is the major source of progesterone throughout pregnancy (Linzell & Heap, 1968; Thorburn & Schneider, 1972). The sudden fall in progesterone concentration before parturition in the goat is the result of regression of the corpus luteum. It follows, therefore, that any effect of foetal corticosteroids is less direct, their action being ultimately on progesterone secretion by the maternal ovary.

Because of their luteolytic effects during the ovine oestrous cycle

[95]

(McCracken, Baird & Goding, 1971), and in other species near parturition, it seems possible that oestrogens or prostaglandins may be intermediary components of the effects of foetal corticosteroids on progesterone secretion near parturition. A striking increase in the concentration of unconjugated oestrogen is seen in maternal jugular plasma on the day of parturition in sheep (Challis, 1971; Thorburn *et al.* 1972) and there is little doubt that this oestrogen is of foeto-placental origin although Thorburn *et al.* (1972) found no detectable unconjugated oestrogens in foetal plasma at this time. However, high oestrogen sulphoconjugate concentrations occur in foetal plasma (Findlay & Cox, 1970), probably as a result of the preponderance of sulphotransferase relative to sulphatase activities in placental and foetal tissues (Findlay, 1970; Findlay & Seamark, 1972). Any change in oestrogen biosynthesis by the ovine foeto-placental unit near parturition is therefore likely to be seen in the sulphoconjugate fraction in foetal plasma.

The levels of unconjugated oestrogens in goat plasma are much higher than those of sheep but the relative increase (two- to threefold) on the day of parturition is less marked (Challis & Linzell, 1971; Thorburn *et al.* 1972). The proportions of individual oestrogens present in goat plasma differ from those in the sheep: oestradiol-17α and oestrone are the major compounds measured in goat plasma whereas oestrone and oestradiol-17β predominate in the maternal plasma of the sheep. Significant increases in unconjugated oestrogen levels have been measured in foetal goat plasma, paralleling the changes reported in maternal plasma (Thorburn *et al.* 1972; Shutt, Cox & Thorburn, 1973). Values for the levels of oestrogen sulphoconjugates in goat plasma have not been reported.

Preliminary observations on the concentration of the F group of prostaglandins (PGF) in the utero-ovarian veins of parturient ewes (Liggins & Grieves, 1971; Thorburn *et al.* 1972) provided no evidence for the possibility that PGF might be an intermediate in the action of foetal corticosteroids on placental progesterone secretion. In fact the results suggested that changes in the concentration of PGF were more closely related in time to the changes in oestrogen levels described above. The concentrations of PGF in utero-ovarian venous plasma of the goat also increased on the day of parturition (Thorburn *et al.* 1972) but no appreciable change was observed on the day on which regression of the corpus luteum occurred. However, the authors were aware of the limitations inherent in sampling only once daily and noted the possibility that luteolysis might be effected by acute releases of PGF.

The present experiments were undertaken to study in greater detail the changes occurring immediately before parturition especially in the plasma concentrations of PGF and oestrogens. Blood samples were obtained frequently near parturition in an attempt to define the temporal relationships between changes in the various hormones measured. Because of their likely importance in reflecting oestrogen biosynthesis in the foeto-placental unit, the plasma levels of oestrogen sulphoconjugates in foetal sheep and goats were measured near parturition. Furthermore, interest was focused on the changes in oestrogens and other hormones during the induction of parturition by infusion of foetal sheep with dexamethasone, in contrast to adrenocorticotrophin (ACTH) which had been used in the earlier study (Thorburn *et al.* 1972). The effects of such infusions of a synthetic corticosteroid could be of particular significance since the foetal adrenal might be implicated in oestrogen biosynthesis in the ovine foeto-placental unit (Davies, Ryan & Petro, 1970).

METHODS

Animals

All experiments were performed using Merino ewes and Saanen does at known times after fertile mating. Pregnancy was confirmed and the number of foetuses determined by radiography before operations. Animals were housed indoors after surgery and fed a daily ration of 1 kg of lucerne chaff:crushed oats (1:1).

Surgery

Catheters (PVC, 1·0 mm i.d. × 2·0 mm o.d.) were inserted into the carotid artery and jugular vein of foetuses using the procedures described by Bassett, Thorburn & Wallace (1970) and Thorburn *et al.* (1972). Additional PVC catheters (1·0 mm i.d. × 1·5 mm o.d.) were placed in one or both utero-ovarian veins of some of the ewes and both does via a branch of the uterine vein on the lateral margin of the uterus and broad ligament. In one ewe (5284) an additional polythene catheter (2·0 mm i.d. × 3·0 mm o.d.) was placed into the amniotic sac via a separate hysterotomy and securely tied in with a double purse-string suture through the myometrium and membranes.

Infusions of foetuses

The foetuses of three ewes (126, 133 and 206) were infused with dexamethasone beginning on days 117, 119, and 128 of gestation and continuing until delivery. Decadron (dexamethasone phosphate, Merck, Sharpe and Dohme) was diluted in saline containing Terramycin (oxytetracycline

HCl, Pfizer) and infused into the jugular vein at a rate of 50 μg dexamethasone phosphate/h.

Blood sampling

Blood was drawn from the foetal carotid artery and maternal uteroovarian and jugular veins into heparinised tubes and kept on ice for at most 30 min before separating the plasma by centrifugation at 4 °C. Plasma samples were stored at -10 °C until analysis. Samples were taken at least daily after surgery but as parturition approached the frequency was increased to as often as every few minutes as shown in Figs. 1–8.

Amniotic pressure recordings

The large catheter was connected to a pressure transducer (S.E. Laboratories, Middlesex, Type S.E. 4/81 Mk 2) maintaining hydrostatic continuity with the amniotic fluid. The transducer was calibrated against a mercury manometer and recorded continuously on a Devices strip chart recorder.

Hormone assays

Plasma was deproteinised with ethanol or extracted with hexane. Corticosteroids and progesterone were measured against cortisol and progesterone standards using the competitive protein-binding assay of Bassett & Hinks (1969). The validity of using washed hexane extracts for progesterone determination has been demonstrated for ovine (Thorburn & Mattner, 1971) and caprine plasma (Thorburn & Schneider, 1972). The concentration of the F group of prostaglandins was determined by radioimmunoassay after group purification using microcolumn chromatography on silica gel (Thorburn *et al.* 1972; Cox, Schneider & Thorburn, 1973).

Plasma concentrations of oestrone (E_1), oestradiol-17α ($E_{2\alpha}$) and oestradiol-17β ($E_{2\beta}$) were measured by competitive protein binding using ovine uterine cytosol (Shutt & Cox, 1972, 1973; Thorburn *et al.* 1972). Solvolysis of oestrogen sulphoconjugates [E(S)] in ethyl acetate by the method of Burstein & Lieberman (1958), was followed by alkali and water partition, chromatography and assay of the liberated oestrogens as described above. Solvolysis resulted in a mean recovery of 74% $\pm$ 5 (S.D.) ($n = 6$) of 0·5 μCi (41 ng) potassium [6,7-^{3}H]oestrone sulphate being extracted as unconjugated oestrone. Because labelled oestrogen sulphoconjugates of sufficiently high specific activity were not available, procedural losses during assay were estimated using [³H]E$_{2\beta}$ and the results have been corrected for such losses. The mean recovery of [³H]E$_{2\beta}$ added to plasma and taken through the method was 69% ±8 (S.D.) (n = 63).

RESULTS

Changes in foetal and maternal oestrogens and maternal prostaglandin F in ewes before spontaneous parturition at term

Four of the ewes were studied throughout late pregnancy and during spontaneous parturition at term (150 days). The utero-ovarian veins of ewes 2924 and 3969 were not catheterised and the data presented are restricted to changes in foetal and maternal oestrogen concentrations. Ewes 5284 and 6522 were sampled more intensively before delivery and provided detailed information on hormone changes at this time.

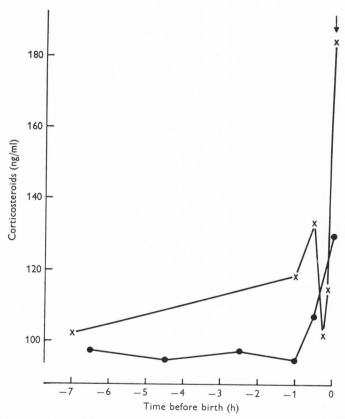

Fig. 1. Plasma corticosteroid concentrations in the foetuses of two ewes (5284, × ; 6522, ●) before and during parturition. The arrow indicates the time of delivery.

Progesterone concentrations in ewes 5284 and 6522 at parturition had fallen to less than 3 ng/ml in the jugular vein and about 10 ng/ml in the utero-ovarian vein ipsilateral to the pregnant horn. There was a close parallel between changes in the two veins with an approximately fivefold difference in concentration. The level of corticosteroids in the foetal plasma in ewe 5284 rose during the last 10 days to reach very high concentrations at birth. Frequent sampling during labour indicated rapid changes between 100 ng/ml and 185 ng/ml (Fig. 1). The highest concentrations were measured in the samples taken during vaginal passage (185 and 130 ng/ml in the foetuses of ewes 5284 and 6522, respectively).

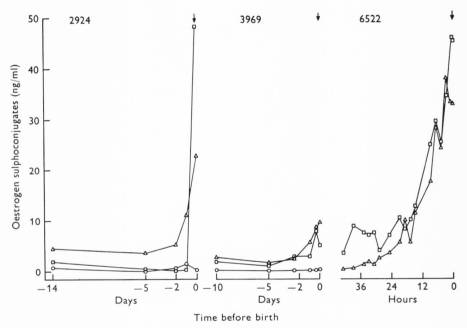

Fig. 2. Plasma concentrations of oestrogen sulphoconjugates in three foetuses (ewes 2924, 3969 and 6522). Sulphoconjugated oestrone (△); sulphoconjugated oestradiol-17α (□); sulphoconjugated oestradiol-17β (○). Note differences in the time scale. The arrows indicate time of delivery.

Samples obtained from the foetuses of two ewes (2924, 3969) between days 136 and 148 indicated that sulphoconjugated oestrone $[E_1(S)]$ and sulphoconjugated oestradiol-17α $[E_{2\alpha}(S)]$ were present in concentrations of 2–4 ng/ml whereas the concentrations of sulphoconjugated oestradiol-17β $[E_{2\beta}(S)]$ were very much lower. Concentrations increased on day 149 (3–11 ng/ml) and large increases in $E_1(S)$ and $E_{2\alpha}(S)$ and a small increase in $E_{2\beta}(S)$ were evident on the day of parturition (Fig. 2). In a later experi-

ment, the foetus of ewe 6522 was sampled frequently during the last 40 h
in order to measure increases in $E_1(S)$ and $E_{2\alpha}(S)$ before parturition in
greater detail. The concentrations increased steadily from 0·5 ng/ml and
3·5 ng/ml, respectively at −40 h to peak levels of 38 ng/ml and 46 ng/ml
shortly before delivery.

Of the two major $E(S)$ components measured in foetal plasma, the
concentration of $E_1(S)$ was marginally the higher during the last 2 weeks
of gestation although some variability was observed on the last day (Fig. 2).

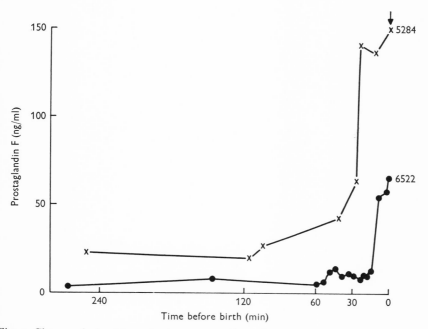

Fig. 3. Changes during labour in the concentrations of prostaglandin F in utero-ovarian
venous plasma draining the pregnant horn in each of two ewes (5284, ×; 6522, ●).
The arrow indicates the time of delivery.

The sulphoconjugates are the predominant form of the oestrogens in
foetal plasma. Thorburn *et al.* (1972) and Shutt *et al.* (1973) showed that
unconjugated oestrogen levels were undetectable (below 40 pg/ml) even
6 h before parturition. However, as seen in Fig. 2, the sulphoconjugates
were measurable in ng/ml and thus the ratio of sulphoconjugated oestro-
gen:unconjugated oestrogen [E(S):E] was very high being over 10^2 and
possibly as high as 10^3. Results obtained from maternal plasma (Cox,
Wong, Currie & Thorburn, 1972) indicate ratios of E(S):E ranging from
3 to 60 and contrast markedly with the foetal ratios. Since the foetal oestro-

gens are almost entirely in the sulphoconjugate form, further assays were confined to this fraction.

The levels of PGF in utero-ovarian venous plasma remained low (< 2·5 ng/ml) until about 25 h before parturition. Concentrations increased gradually in ewe 6522 with a slight rise during the first stages of labour but no sizeable change was observed until the beginning of second stage labour (last 15 min, Fig. 3). The changes in ewe 5284 differed in that a broad peak of some 23 h duration and reaching 32 ng/ml was observed (see Fig. 5). This was followed by a more rapid rise (20–63 ng/ml) during the first stage of

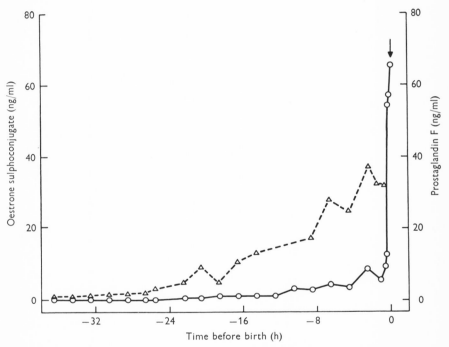

Fig. 4. Comparison between changes in foetal sulphoconjugated oestrone (△) and maternal utero-ovarian venous prostaglandin F (○) concentrations in ewe 6522. The arrow indicates the time of delivery.

labour and a very striking increase at the onset of the second stage (indicated by the presentation of the head and forelimbs in the cervix) 25 min before delivery was completed (Fig. 3). In both animals, the interval during which the greatest increase occurred was 4 min and was accompanied by vigorous straining efforts on the part of the ewe. A continuous recording of amniotic pressure changes in ewe 5284 showed that during the last 120 min regular contractions occurred.

It has been established that maternal unconjugated oestrogen concentrations show a rise which is very similar to those in foetal E(S) indicated in Fig. 2. Thus the relationship between changes in foetal $E_1(S)$ and maternal utero-ovarian venous PGF concentrations for ewe 6522 were examined (Fig. 4). For clarity, foetal $E_{2\alpha}(S)$ data (see Fig. 2) have been omitted. It is clear that the increases in foetal E(S) preceded the main changes in utero-ovarian PGF, although the first samples in which PGF concentration could be distinguished from blank values (-22 h) corresponded approximately to the start of the final increase in foetal E(S) levels.

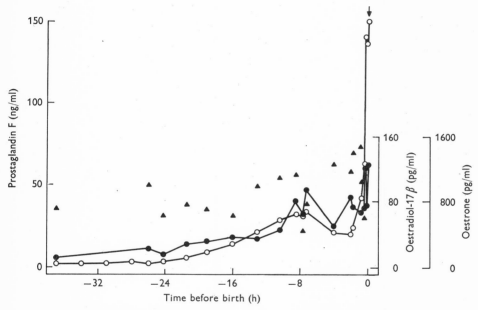

Fig. 5. Relationship between concentrations of prostaglandin F (○) and unconjugated oestrone (▲) and oestradiol-17β (●) in the utero-ovarian venous plasma of ewe 5284 before delivery. The arrow indicates time of delivery.

Inadequate foetal sampling during labour prevented a close examination of the relationship when PGF levels were changing rapidly. Since the initial PGF changes in ewe 5284 were slower, the relationship with unconjugated oestrogen concentrations could be examined more readily. Concentrations of E_1, $E_{2\beta}$ and PGF were measured in utero-ovarian venous plasma during the 38 h before delivery and are presented together in Fig. 5. The levels of $E_{2\beta}$ and PGF changed in a very similar fashion during this period although it was apparent that all three increased. However, the concentration of PGF rose disproportionately over the last few minutes (Fig. 5).

Comparative hormonal changes in goats delivering at term

Observations made on two goats which delivered single foetuses at term (150 days) have been included for comparative purposes. After catheterisation on days 121 and 129, respectively, goats 241 and 155 were sampled daily until day 144 and more frequently thereafter. Corticosteroid levels in foetal plasma rose sharply during the last 4–5 days, but in both foetuses the increase began about 14 days before birth. The very high concentrations (400 ng/ml) in one foetus during vaginal passage are shown in Fig. 6.

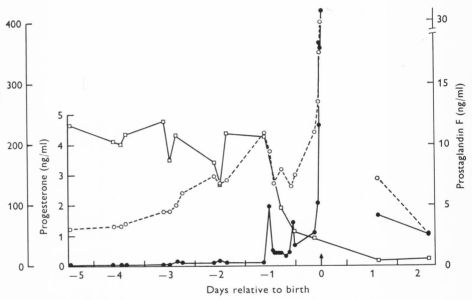

Fig. 6. Plasma hormone changes in goat 241 around parturition. Foetal corticosteroid concentration (○) maternal jugular progesterone (□) and maternal utero-ovarian venous prostaglandin F (●). Oestrogen changes are shown in Fig. 7. The arrow indicates the time of birth.

Peripheral progesterone concentrations in both goats fell suddenly about 24 h before delivery and were below 1 ng/ml at the onset of labour. The fall in progesterone in goat 241 coincided with an acute peak of PGF in the utero-ovarian vein draining the uterine horn ipsilateral to the ovary which contained the corpus luteum (Fig. 6). A further brief secretion of PGF was detected some 12 h before delivery and a major increase occurred during labour. The greatest change was seen during the early part of second stage labour when the cervix was fully dilated and the highest detected concentrations of PGF (31·5 ng/ml) was in the sample taken 1 min after the completion of delivery. Similar measurements could not be made in goat 155 because the utero-ovarian vein catheter was no longer patent after surgery.

Changes in foetal and maternal E(S) and E concentrations are illustrated in Fig. 7. Results are shown for sulphoconjugated and unconjugated E_1 and $E_{2\alpha}$ components only, since the $E_{2\beta}$ components were undetectable (below 10 per cent of the $E_{2\alpha}$ level) in both foetal and maternal goat plasma. Both $E_1(S)$ and $E_{2\alpha}(S)$ were present in foetal plasma at about 10 ng/ml on day 137 and increased gradually to concentrations of up to

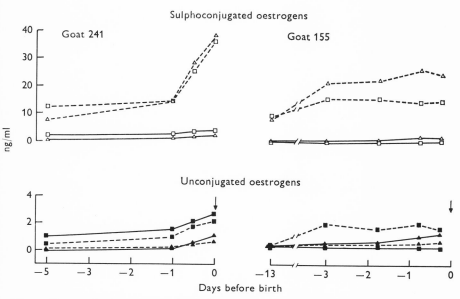

Fig. 7. Changes in the concentrations of unconjugated and sulphoconjugated oestrogens in the foetal (− − − −) and maternal (———) plasma of two goats before spontaneous parturition at term. The upper half of the figure shows sulphoconjugated oestrone (△) and sulphoconjugated oestradiol-17α (□); the lower half, unconjugated oestrone (▲) and unconjugated oestradiol-17α (■). The arrows indicate the time of delivery.

22 ng/ml by day 147. Further increases in E(S) were less marked than in foetal sheep during the last few days of gestation. A smaller, but quite obvious increase in E_1 and $E_{2\alpha}$ concentration in foetal plasma paralleled changes in the concentration of E(S). The ratios of E(S):E (Table 1) are much smaller than in the sheep foetus and do not show a consistent pattern of change before parturition. Analyses of maternal plasma for both conjugated and unconjugated fractions confirmed the changes in unconjugated E_1 and $E_{2\alpha}$ previously reported (Thorburn *et al.* 1972) and indicated that a similar increase in $E_1(S)$ and $E_{2\alpha}(S)$ occurred before parturition. Maternal unconjugated oestrogen concentrations in the goat were higher

Table 1. *Ratio of sulphoconjugated to unconjugated oestrogens in two goats and their foetuses in late gestation*

	Time before parturition (h)	Ratio of sulphoconjugated:unconjugated oestrogen			
		Foetal plasma		Maternal plasma	
		Oestrone	Oestradiol-17α	Oestrone	Oestradiol-17α
Goat 155	312	19	20	2	1
	72	47	8	4	2
	42	39	10	4	4
	18	44	7	7	5
	6	30	9	4	5
Goat 241	96	26	31	—	2
	24	70	15	3	2
	12	65	15	3	2
	0·5	71	18	2	2

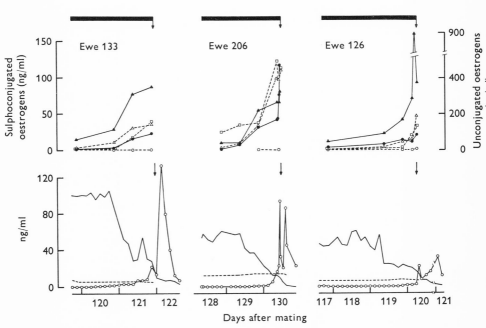

Fig. 8. Plasma hormone changes in three sheep during the induction of parturition by infusing the foetuses with dexamethasone phosphate. Infusions (shown by black horizontal bars at the top of the figure) continued until premature delivery (indicated by arrow). The top half of the figure shows foetal sulphoconjugated oestrogens: oestrone (△– – – –△), oestradiol-17α (□– – – –□), oestradiol-17β (○– – – –○); unconjugated oestrogens: oestrone (▲——▲), oestradiol-17β (●——●). The bottom half of the figure shows foetal corticosteroids (– – – –), maternal utero-ovarian venous progesterone (———) and prostaglandin F (○——○).

than those in sheep at comparable stages of gestation but the rise immediately before parturition was less marked. It can be seen from Fig. 7 and Table 1 that the maternal E(S):E ratios were smaller than in foetal plasma but were similar to values obtained in ovine maternal plasma.

Premature parturition in sheep: dexamethasone-treated foetuses

Parturition was induced in three ewes (126, 133 and 206) after infusing their foetuses with dexamethasone phosphate (50 μg/h) for 62, 53 and 50 h, respectively. The synthetic steroid infused did not interfere in the corticosteroid assay used and parturition was accomplished without increases of endogenous foetal corticosteroids (Fig. 8). There was no evidence of a decrease in endogenous corticosteroids during the infusion and there were also no signs of acute corticosteroid responses after birth despite the hypoxic stress associated with respiratory failure.

The changes observed in maternal progesterone levels varied somewhat between ewes. Jugular and utero-ovarian concentrations were again generally in the ratio of 1:5 and changes in the utero-ovarian vein were reflected in more gradual changes in the jugular vein. The fall in progesterone levels some 24–30 h after beginning the foetal infusion was striking in ewe 133, less so in ewe 206 and least pronounced in ewe 126. The interval between surgery and the beginning of infusion in ewe 133 was short so that the marked decline in progesterone concentration in this ewe may have been because pre-infusion levels were spuriously high. After the first 30 h, the fall became more gradual so that jugular progesterone concentrations were 6·5, 3·6 and 3·8 ng/ml just before labour in ewes 126, 133 and 206, respectively. An abrupt decline in utero-ovarian venous progesterone occurred after delivery and placental separation in ewe 206 (Fig. 8) but in the others, where delivery of the placentae was somewhat delayed, this post-partum fall was more gradual.

The concentrations of $E_1(S)$ and $E_{2\alpha}(S)$ in foetal plasma increased about 24 h after the start of infusions but the greatest increases were observed during the 24 h before delivery. The pattern of change in E(S) levels was similar to those of full-term foetuses (Fig. 2) in both the relative and absolute amounts of $E_1(S)$ and $E_{2\beta}(S)$ measured. Concentrations of $E_{2\beta}(S)$ remained low throughout the infusion. The concentrations of maternal E_1 and $E_{2\beta}$ were much lower than the concentrations of sulphoconjugated oestrogens in foetal plasma (about 1–2 per cent). The increases during the last 24 h before delivery generally mirrored changes in the concentrations of foetal hormones both in timing and the relative change except for the very high E_1 level in ewe 126 shortly before parturition.

The concentration of PGF in the utero-ovarian veins remained low until shortly before birth. A very slight increase was apparent in the utero-ovarian veins of ewe 126 some 9 h before parturition and increased concentrations (22 ng/ml) were observed immediately after delivery. These had fallen in the next sample but then a second more gradual rise was observed such that 32 ng/ml were still present 27 h after birth. A slow rise in PGF levels in the vein draining the pregnant horn of ewe 133 was observed during the 30 h preceding parturition with an accelerated increase being noted during the last 9 h. Although no samples were taken at birth and its exact timing was unknown, the appearance of the ewe and its dead newborn indicated that it had delivered close to the time shown by the end of the infusion in Fig. 8. The first sample taken after birth contained 13 ng/ml and thereafter the concentration increased for 12 h when a peak concentration of 133 ng/ml was observed. Evidence of further changes in PGF concentration up to 40 h after birth was obtained. PGF levels in the utero-ovarian veins of ewe 206 began to rise 7 h before parturition and thereafter rose in a steady and finally rapid fashion to reach levels of 94 ng/ml in the vein draining the pregnant horn and 44 ng/ml in the contralateral vein. Figure 8 shows that peak concentrations at parturition were quite acute but, as in the case of ewe 133, were followed by more gradual increases after birth. Levels were still high (21–24 ng/ml) 21 h after birth. Of the three ewes, only one (206) showed a large acute increase in PGF concentration similar to those shown in Fig. 3 for ewes delivering at term. This ewe differed from the others in the pattern of increase in maternal $E_{2\beta}$ concentrations and the final increases in $E_{2\beta}$ and PGF occurred simultaneously (Fig. 8).

DISCUSSION

The foetal adrenal cortex and initiation of parturition

The role of the foetal hypothalamus, pituitary and adrenal cortex in initiating the endocrine changes in both the foetus and the mother which lead to parturition have been discussed by Liggins (1969), Liggins, Grieves, Kendall & Knox (1972), Thorburn *et al.* (1972) and Bassett & Thorburn (1972). The last two papers have demonstrated the likelihood of a similar mechanism in the goat by inducing premature parturition with infusion of ACTH into the foetus. In the present study plasma concentrations of corticosteroids in foetal goats rose sharply before spontaneous parturition at term in a manner similar to that described in the foetal lamb (Bassett & Thorburn, 1969, 1972). Corticosteroid concentrations in three foetal goats were highest during delivery but fell subsequently as was found in the lamb by Bassett & Alexander (1971). We have extended previous observations on corticosteroid levels in sheep foetuses at term by demonstrating rapid changes in concentration during labour. These may well reflect responses to stresses imposed on the foetus at this time by uterine contractions and partial asphyxia (Comline & Silver, 1972). The latter is a potent stimulus for ACTH secretion in the ovine foetus (Alexander *et al.* 1971) and the adrenal cortex of the term foetus is sufficiently sensitive to ACTH (Bassett & Thorburn, 1972; Madill & Bassett, 1973) for plasma corticosteroid levels to reflect acute changes in ACTH secretion. Similar acute changes in foetal corticosteroids have been seen during labour in preparations in which foetuses were infused with ACTH at a constant rate (W. B. Currie & G. D. Thorburn, unpublished). As expected, there were no such changes in corticosteroid levels in the dexamethasone-infused foetuses. Since the adrenal cortices of foetuses of 120–130 days of gestation are likely to be insensitive to acute changes in endogenous ACTH, we cannot comment on the possible overriding effect of stresses associated with delivery on the likely suppression of ACTH secretion by the infused dexamethasone. Neonates born prematurely after dexamethasone infusion all died within 36 h but there was no evidence of any increase in plasma corticosteroids despite the development of terminal respiratory failure. The premature neonates were studied in a collaborative project with Dr G. Alexander and will not be considered further in this paper.

Progesterone changes in ewes and goats before parturition

Progesterone concentrations in the peripheral plasma of ewes fall before parturition at term (Bassett *et al.* 1969; Fylling, 1970) and similar changes precede parturition induced prematurely by foetal infusion with ACTH (Bassett & Thorburn, 1972). Liggins (1969) made three measurements of peripheral progesterone levels before and during the course of dexamethasone-induced parturition in a single ewe and found a similar change. In the present study we have examined the time-course of these changes in greater detail in an attempt to relate them to the rapidly altering endocrine profiles of both foetus and mother. Since maternal progesterone levels decreased some time after foetal corticosteroid concentrations had risen in response to endogenous ACTH secretion near term, or in response to infusion of the foetus with synthetic glucocorticoids, a causal relationship between the changes in these hormones in foetal and maternal plasma appears likely. This suggestion was first advanced by Liggins (1969) but a full explanation for the mechanisms involved is still lacking. On the basis of kinetic studies, Bedford, Challis, Harrison & Heap (1972) concluded that a decline in progesterone production rate before parturition was largely responsible for the plasma concentration changes observed. Decreased production could result from a drop in placental biosynthesis, increased placental metabolism at the site of synthesis (Ainsworth & Ryan, 1967) or an alteration in the extent of foetal metabolism of the hormone (Nancarrow, 1969). As yet, there is no direct evidence for the most likely possibility – decreased placental biosynthesis – but studies performed *in vitro* by Ainsworth & Ryan (1967) do indicate extensive metabolism of progesterone by the ovine placenta near term.

Factors affecting the sharp decrease in the concentration of progesterone in the circulation before parturition (Thorburn & Schneider, 1972; Thorburn *et al.* 1972) can be examined more readily in the goat than in the sheep. The goat is almost entirely dependent upon the corpus luteum for progesterone secretion (Linzell & Heap, 1968; Thorburn & Schneider, 1972) and the change observed just before parturition must result from regression of the corpus luteum. Since this can be brought about by chronic stimulation of the adrenal cortex of the goat foetus either pathologically (van Rensberg, 1971) or experimentally (Thorburn *et al.* 1972), a signal, ultimately derived from the foetus, is effective at the maternal corpus luteum. The sudden decrease in jugular progesterone levels seen in goat 241 some 25 h before delivery, occurred when corticosteroids had reached high levels in the foetal plasma and very close to the sudden rise of PGF levels in the utero-ovarian vein (Fig. 6). This observation supports the suggestion, albeit then unsubstantiated, of Thorburn *et al.* (1972) that $PGF_{2\alpha}$ could be a component of the proposed signal. Indeed, brief infusions of $PGF_{2\alpha}$ into a uterine vein ipsilateral to the ovary containing the corpus luteum in pregnant does caused rapid luteolysis and premature parturition occurred about 30 h later (Currie & Thorburn, 1973). The origin of the prostaglandin measured in the utero-ovarian veins on day 149 and the factors controlling its release are not known but its origin is likely to be placental; perhaps it is released in response to chronically increased corticosteroid or oestrogen levels in foetal plasma. It should be noted that this release occurred when the uterus and placenta were exposed to levels of progesterone typical of pregnancy whereas labour and the associated massive increase in utero-ovarian venous PGF both follow a withdrawal of progesterone (see below). Further experiments are in progress to examine the control of the luteolytic secretion of $PGF_{2\alpha}$ in the goat and the mechanism of corpus luteum regression that follows.

Foetal and maternal oestrogens

The data obtained from foetal sheep in the present study clearly show a massive increase in the concentrations of E(S) on the day of parturition. This increase parallels the changes in maternal levels of unconjugated oestrogens already established (Challis, 1971; Challis, Harrison & Heap, 1971; Bedford *et al.* 1972; Thorburn *et al.* 1972; Shutt *et al.* 1973) and suggests that foeto-placental biosynthesis of oestrogen increases at this time with consequent transfer of E or E(S) to the ewe's circulation. Oestrogens are synthesised in the foetal component of the ovine placenta, and C_{19} steroids can act as substrates (Ainsworth & Ryan, 1966; Findlay, 1970; Pierrepoint, Anderson, Harvey, Turnbull & Griffiths, 1971; Findlay & Seamark, 1972). Transfer of oestrogen from the foetal to the maternal circulation has been demonstrated with labelled E_1 (Findlay & Seamark, 1972) and $E_1(S)$ (M. S. F. Wong, R. I. Cox, W. B. Currie & G. D. Thorburn, unpublished). The preponderance of sulphotransferase over sulphatase activities in the foeto-placental tissues results in the rapid sulphoconjugation of oestrogens (Findlay, 1970; Findlay & Seamark, 1972), and thus the predominance of E(S) over E in foetal plasma (Findlay & Cox, 1970). The absence of measurable unconjugated oestrogens in foetal plasma before parturition (Thorburn *et al.* 1972; Shutt *et al.* 1973) compared with the changes in E(S) presented here undoubtedly reflects the disparate extent of sulphoconjugation. The high concentrations of E(S) from chronically catheterised preparations, measured by protein-binding assays, are in general agreement with those based on spectrophoto-fluorimetric analyses of samples obtained by cardiac puncture of foetal sheep at slaughter (Findlay & Cox, 1970).

The precise mechanism by which the surge in maternal unconjugated oestrogens occurs on the day of parturition has yet to be elucidated. The increase in production rate is undoubtedly of foeto-placental origin since a similar change has been observed in the ovariectomised ewe (Bedford *et al.* 1972). The form in which oestrogen crosses to the maternal circulation is unknown but a considerable proportion may be as sulphoconjugates which are subsequently re-equilibrated in the maternal tissues to give the reduction in E(S):E ratio in peripheral plasma (Thorburn *et al.* 1972). Another aspect is the different proportions of oestrogens in foetal and maternal plasma. In the foetus, E_1 and $E_{2\alpha}$ are quantitatively the major oestrogens (Fig. 2) and are almost entirely sulphoconjugated. In maternal plasma, E_1 and $E_{2\beta}$ are the two main oestrogens with a much higher proportion being unconjugated. A better understanding of these changes should be gained from further experiments using radioactive oestrogens.

The increases in foetal E(S) concentrations observed during the infusion of dexamethasone are of considerable interest. We have found increases in both foetal E(S) and maternal E in all three animals in which parturition was induced with dexamethasone by 130 days. This contrasts with the equivocal changes in maternal $E_{2\beta}$ reported by Liggins *et al.* (1972) in sheep induced earlier than 130 days. Since changes in foetal E(S) and maternal E were comparable to those in animals at term, it seems certain that the provision of precursors to the placenta was adequate (cf. Anderson, Pierrepoint, Griffiths & Turnbull, 1972) despite the immaturity of the foetus. The ovine foetal adrenal cortex has been implicated in studies *in vitro* (Davies *et al.* 1970; Pierrepoint *et al.* 1971) as a source of C_{19} precursors for placental aromatisation to oestrogen, but its contribution *in vivo* has not been assessed. Since the foetal pools of androstenedione and dehydroepiandrosterone sulphate [DHEA(S)], are turned over quite rapidly (Findlay, 1970), any marked increase in foeto-placental biosynthesis of E(S) or E would require an increased provision of precursors. Since endogenous corticosteroid concentrations in foetal plasma did not change during dexamethasone infusion, it appears that precursors for placental aromatisation are made available through pathways not closely related to those of corticosteroid synthesis.

Thus it may be speculated that dexamethasone, by an effect on steroid bio-synthesis, increases the supply of C_{19} precursors for placental oestrogen biosynthesis; alternatively, there may be more direct effects on oestrogen biosynthesis in the placenta. It appears that the effect of dexamethasone on placental progesterone biosynthesis is similar to that of endogenous corticosteroids, so that the latter may also be directly responsible for some of the changes in oestrogen biosynthesis occurring before parturition at term. These associations between corticosteroids in the foetus and oestrogen biosynthesis re-emphasise the possible importance of oestrogens in the control of parturition. In support of this was the finding that three out of four ewes infused with physiological amounts of $E_{2\beta}$ at 133 days of gestation went into premature labour and delivered their foetuses within 36 h. Androstenedione infused into a foetus in one experiment also resulted in premature parturition, whereas infusion of similar amounts of DHEA(S) was ineffective (W. B. Currie, M. S. F. Wong, R. I. Cox & G. D. Thorburn, unpublished).

The changes in oestrogen concentration in foetal and maternal goat plasma are less complex than those in the sheep. The concentrations of 'total' oestrogens in maternal plasma (Challis & Linzell, 1971) are much higher in the goat throughout pregnancy. Thorburn *et al.* (1972) and Shutt *et al.* (1973) measured E_1 and $E_{2\alpha}$ individually and found high levels and similar changes in both foetal and maternal plasma during the last two weeks of pregnancy. There was less similarity in the concentrations of E(S) in foetal and maternal plasma (Fig. 7). The concentrations of $E_1(S)$

and $E_{2\alpha}(S)$ are higher in foetal plasma than in the plasma of the doe. The proportion as E(S) in the goat foetus is very much less than was found in foetal sheep but in both species a major increase occurs in the E(S) fraction before parturition. Whilst oestrogen concentrations in the pregnant goat are higher than in sheep, the relative increase shortly before parturition is less than in the sheep.

Prostaglandin F

The measurement of PGF in utero-ovarian venous plasma of three ewes at term showed that there were two main phases of increase (Figs. 3 and 4). PGF was present at only low concentrations until about 24 h before parturition. However, 4–10 h before parturition some increase in PGF concentrations could be seen, particularly in ewe 5284. The major increase was very rapid and occurred principally during the last 25 min before birth. These very rapid increases in PGF probably account for the somewhat lower levels measured near parturition by Thorburn *et al.* (1972) in ewes where the interval between sampling and delivery was not known so accurately. In the case of dexamethasone-induced parturition in three ewes, the pattern of change in PGF concentrations was essentially the same (Fig. 8). Again, the relatively small increases in PGF seen in ewes 126 and 133 probably resulted from sampling with too low a frequency in the period just before delivery. The physiological significance of acute changes in PGF concentrations in the goat on the day preceding parturition (Fig. 6) has already been discussed and a hypothesis advanced to account for PGF release at that time.

More striking changes in PGF levels in utero-ovarian venous plasma occurred in the goat and ewes during labour. In one animal (ewe 5284) the gradual increase in PGF over the last 24 h occurred whilst the amniotic pressure showed wavelike changes suggestive of the onset of synchronised uterine contractions. One of the recognised interactions between prostaglandins and the myometrium is an improved co-ordination of contractions which Pickles (1967) considered to reflect better conduction of excitatory processes from cell to cell. The most obvious increase in PGF levels occurred over a very short interval corresponding to the presentation of the forelimbs and head of the foetus in the cervix. At this time the pressure changes took on the form of second stage labour and included a component resulting from abdominal straining closely synchronised with the myometrial contractions as described by Hindson, Schofield, Turner & Wolff (1965). Since this is when oxytocin secretion occurs (Fitzpatrick, 1961), the possibility of the prostaglandins being released as a consequence of enhanced uterine contractions cannot be ignored. Liggins & Grieves (1971) have shown that the concentrations of $PGF_{2\alpha}$ in maternal cotyledons and myometrium after 48 h of foetal infusion with dexamethasone are not affected by progesterone given to the ewes in doses sufficient to suppress labour. We have performed similar experiments in goats (W. B. Currie & G. D. Thorburn, unpublished) in which parturition induced by foetal infusion of ACTH can be completely blocked by injecting the doe with physiological doses (20 mg/day) of progesterone (Thorburn *et al.* 1972). There was a complete absence of change in PGF levels in the utero-ovarian vein until progesterone treatment was withdrawn and labour was allowed to proceed. Thus, whilst the appearance of $PGF_{2\alpha}$ in tissues appears not to be dependent upon labour, there remains the possibility that the release of PGF into the blood could be a consequence of enhanced myometrial activity.

An alternative explanation for the increased concentrations of PGF during labour is their possible relationship with changing concentrations of oestrogens observed at that time (Figs. 4, 5, 8). Unconjugated oestrogens, particularly $E_{2\beta}$, may stimulate synthesis and release of PGF from the myometrium in a way similar to that demonstrated in non-pregnant guinea-pigs by Blatchley *et al.* (1971).

Further discussion of the control of the sudden increase in PGF levels during labour must be speculative until we establish whether the PGF being measured in utero-ovarian venous plasma is of placental or myometrial origin. High concentrations of $PGF_{2\alpha}$ in the maternal cotyledons (Liggins *et al.* 1972) and the large proportion of uterine blood flow being directed to the placenta (Makowski, Meschia, Droegemueller & Battaglia, 1968) suggest a greater contribution of PGF from the placenta. However, the striking increase in utero-ovarian blood flow on the day of parturition (Bedford *et al.* 1972) is likely to be almost entirely from the myometrium and is probably due to the specific hyperaemic effects of the raised levels of oestrogen. Increased myometrial flow at that time would be physiologically consistent with a greater contribution of PGF from the myometrium.

CONCLUSION

The frequent sampling used in the present study has helped to clarify the temporal relationships among a number of hormone changes occurring in both foetuses and dams before parturition. In sheep and goats, increased concentrations of corticosteroids in foetal plasma precede all the major hormone changes seen in the maternal plasma. The raised corticosteroid levels appear to be responsible for the decline in maternal progesterone concentrations before parturition although the mechanisms by which this is achieved differ in the two species. In the sheep there may be a direct interference with placental progesterone synthesis and secretion whereas in the goat the release of PGF into the maternal uterine veins provides a luteolytic signal with subsequent prompt and complete withdrawal of progesterone.

The early changes in plasma concentrations of unconjugated oestrogens and PGF occur over a period corresponding to the duration of labour in the ewe (Hindson *et al.* 1965; Hindson, Schofield & Turner, 1968). If the utero-ovarian vein concentrations of PGF are related to the levels in the myometrium, the increases occurring during labour probably reflect the synthesis of PGF in response to a requirement in connection with uterine contraction. Should the trigger for myometrial PGF synthesis prove to be oestrogen, our finding of close association between cortico-steroids in the foetus and oestrogen biosynthesis by the foeto-placental unit assumes greater importance in the chain of events linking the foetus with the initiation of parturition.

Although much higher concentrations of oestrogen are present through-out pregnancy in the goat (Challis & Linzell, 1971), the increase at parturi-tion relative to levels on the previous day is less striking than it is in sheep. The abrupt and complete withdrawal of progesterone in the goat, when considered alongside the oestrogen changes, gives rise to a rapidly increas-ing oestrogen:progesterone ratio, very similar to that found in the ewe on the day of parturition. Our observations in both species are therefore consistent with the notion of an increased influence of oestrogen relative to progesterone being the trigger for the onset of labour.

ACKNOWLEDGEMENTS

The authors thank Miss D. H. Nicol and Messrs W. Schneider and J. A. Avenell for their participation in various phases of this work. Dr D. A. Shutt kindly provided unconjugated oestrogen data for the goat samples. Prostaglandin $F_{2\alpha}$ was made available by courtesy of The Upjohn Co. (Kalamazoo, USA) and The Ono Pharmaceutical Co. (Osaka, Japan). One of us (W. B. C.) was supported by a CSIRO Postgraduate Scholarship held in the School of Biological Sciences, Macquarie University.

REFERENCES

AINSWORTH, L. & RYAN, K. J. (1966). Steroid transformations by endocrine organs from pregnant animals. I. Estrogen biosynthesis by mammalian placental preparations *in vitro*. *Endocrinology*, **79**, 875–883.

AINSWORTH, L. & RYAN, K. J. (1967). Steroid hormone transformations by endocrine organs from pregnant mammals. II. Formation and metabolism of progesterone by bovine and sheep placental preparations *in vitro*. *Endocrinology*, **81**, 1349–1356.

ALEXANDER, D. P., BRITTON, H. G., FORSLING, M. L., NIXON, D. A. & RATCLIFFE, J. G. (1971). The release of corticotrophin and vasopressin in the foetal sheep in response to haemorrhage. *J. Physiol. Lond.* **213**, 31–32.

ANDERSON, A. B. M., PIERREPOINT, C. G., GRIFFITHS, K. & TURNBULL, A. C. (1972). Steroid metabolism in the adrenals of fetal sheep in relation to natural and corticotrophin-induced parturition. *J. Reprod. Fert.* Suppl. **16**, 25–38.

BASSETT, J. M. & ALEXANDER, G. (1971). Insulin, growth hormone and corticosteroids in neonatal lambs. *Biol. Neonat.* **17**, 112–125.

BASSETT, J. M. & HINKS, N. T. (1969). Micro-determination of corticosteroids in ovine peripheral plasma: effects of venipuncture, corticotrophin, insulin and glucose. *J. Endocr.* **44**, 387–403.

BASSETT, J. M., OXBORROW, T. J., SMITH, I. D. & THORBURN, G. D. (1969). The concentration of progesterone in the peripheral plasma of the pregnant ewe. *J. Endocr.* **45**, 449–457.

BASSETT, J. M. & THORBURN, G. D. (1969). Foetal plasma corticosteroids and the initiation of parturition in the sheep. *J. Endocr.* **44**, 285–286.

BASSETT, J. M. & THORBURN, G. D. (1972). Circulating levels of progesterone and corticosteroids in the pregnant ewe and its foetus. In *Endocrinology of Pregnancy and Parturition. Studies in the Sheep*, ed. C. G. Pierrepoint, pp. 126–140. Cardiff: Alpha Omega Alpha Publishing.

BASSETT, J. M., THORBURN, G. D. & WALLACE, A. L. C. (1970). The plasma growth hormone concentration of the foetal lamb. *J. Endocr.* **48**, 251–263.

BEDFORD, C. A., CHALLIS, J. R. G., HARRISON, F. A. & HEAP, R. B. (1972). The role of oestrogens and progesterone in the onset of parturition in various species. *J. Reprod. Fert.* Suppl. **16**, 1–24.

BLATCHLEY, F. R., DONOVAN, B. T., POYSER, N. L., HORTON, E. W., THOMPSON, C. J. & LOS, M. (1971). Identification of prostaglandin $F_{2\alpha}$ in the utero-ovarian blood of guinea-pig after treatment with oestrogen. *Nature, Lond.* **230**, 243–244.

BURSTEIN, S. & LIEBERMAN, S. (1958). Hydrolysis of ketosteroid hydrogen sulfates by solvolysis procedures. *J. biol. Chem.* **233**, 331–335.

CHALLIS, J. R. G. (1971). Sharp increase in free circulating oestrogens immediately before parturition in sheep. *Nature, Lond.* **229**, 208–209.

CHALLIS, J. R. G., HARRISON, F. A. & HEAP, R. B. (1971). Uterine production of oestrogens and progesterone at parturition in the sheep. *J. Reprod. Fert.* **25**, 306–307.

CHALLIS, J. R. G. & LINZELL, J. L. (1971). The concentration of total unconjugated oestrogens in the plasma of pregnant goats. *J. Reprod. Fert.* **26**, 401–404.

COMLINE, R. S., NATHANIELSZ, P. W., PAISEY, R. B. & SILVER, M. (1970). Cortisol turnover in the sheep foetus immediately prior to parturition. *J. Physiol., Lond.* **210**, 141–142.

COMLINE, R. S. & SILVER, M. (1972). The composition of foetal and maternal blood during parturition in the ewe. *J. Physiol., Lond.* **222**, 233–256.

COX, R. I., SCHNEIDER, W. & THORBURN, G. D. (1973). The radioimmunoassay of the prostaglandin F group in the plasma of different species. *Aust. J. exp. Biol. med. Sci.* (in press).

COX, R. I., WONG, M. S. F., CURRIE, W. B. & THORBURN, G. D. (1972). Changes in oestrogen concentrations in the foetal lamb associated with normal and induced parturition. *Proc. Aust. Physiol. Pharmac. Soc.* **3**, 25–26.

CURRIE, W. B. & THORBURN, G. D. (1973). Induction of premature parturition in goats by prostaglandin $F_{2\alpha}$ administered into the uterine vein. *Prostaglandins* (submitted for publication).

DAVIES, I. J., RYAN, K. J. & PETRO, Z. (1970). Estrogen synthesis by adrenal-placental tissues of the sheep and iris monkey *in vitro*. *Endocrinology*, **86**, 1457–1459.

FINDLAY, J. K. (1970). *Oestrogens in the ovine foeto-placental unit*. Ph.D. thesis, University of Adelaide.

FINDLAY, J. K. & COX, R. I. (1970). Oestrogens in the plasma of the sheep foetus. *J. Endocr*. **46**, 281–282.

FINDLAY, J. K. & SEAMARK, R. F. (1972). The occurrence and metabolism of oestrogens in the sheep foetus and placenta. In *Endocrinology of Pregnancy and Parturition. Studies in the Sheep*, ed. C. G. Pierrepoint, pp. 54–64. Cardiff: Alpha Omega Alpha Publishing.

FITZPATRICK, R. J. (1961). The estimation of small amounts of oxytocin in blood. In *Oxytocin*, eds. R. Caldeyro-Barcia & H. Heller, pp. 358–377. New York: Pergamon Press.

FYLLING, P. (1970). The effect of pregnancy, ovariectomy and parturition on plasma progesterone level in sheep. *Acta endocr., Copenh*. **65**, 273–283.

HINDSON, J. C., SCHOFIELD, B. M., TURNER, C. B. & WOLFF, H. S. (1965). Parturition in the sheep. *J. Physiol., Lond*. **181**, 560–567.

HINDSON, J. C., SCHOFIELD, B. M. & TURNER, C. B. (1968). Parturient pressures in the ovine uterus. *J. Physiol., Lond*. **195**, 19–28.

LIGGINS, G. C. (1968). Premature parturition after infusion of corticotrophin or cortisol into foetal lambs. *J. Endocr*. **42**, 323–329.

LIGGINS, G. C. (1969). The foetal role in the initiation of parturition in the ewe. In *Foetal Autonomy*, eds. G. E. W. Wolstenholme & M. O'Connor, pp. 218–231. London: Churchill.

LIGGINS, G. C. & GRIEVES, S. A. (1971). Possible role for prostaglandin $F_{2\alpha}$ in parturition in sheep. *Nature, Lond*. **232**, 629–631.

LIGGINS, G. C., GRIEVES, S. A., KENDALL, J. Z. & KNOX, B. S. (1972). The physiological roles of progesterone, oestradiol-17β and prostaglandin $F_{2\alpha}$ in the control of ovine parturition. *J. Reprod. Fert*. Suppl. **16**, 85–104.

LIGGINS, G. C., KENNEDY, P. C. & HOLM, L. W. (1967). Failure of initiation of parturition after electrocoagulation of the pituitary of the foetal lamb. *Am. J. Obstet. Gynec*. **98**, 1080–1086.

LINZELL, J. L. & HEAP, R. B. (1968). A comparison of progesterone metabolism in the pregnant sheep and goat: sources of production and an estimation of uptake by some target organs. *J. Endocr*. **41**, 433–438.

McCRACKEN, J. A., BAIRD, D. T. & GODING, J. R. (1971). Factors affecting the secretion of steroids from the transplanted ovary of the sheep. *Recent Prog. Horm. Res*. **27**, 537–582.

MADILL, D. & BASSETT, J. M. (1973). Corticosteroid release by adrenal tissue from foetal and newborn lambs in response to corticotrophin stimulation in a perifusion system *in vitro*. *J. Endocr*. **58**, 75–87.

MAKOWSKI, E. L., MESCHIA, G., DROEGEMUELLER, W. & BATTAGLIA, F. C. (1968). Distribution of uterine blood flow in the pregnant sheep. *Am. J. Obstet. Gynec*. **101**, 409–412.

NANCARROW, C. D. (1969). *Gestagen metabolism in the ovine foeto-placental unit*. Ph.D. thesis, University of Adelaide.

PICKLES, V. R. (1967). The prostaglandins. *Biol. Rev*. **42**, 614–652.

PIERREPOINT, C. G., ANDERSON, A. B. M., HARVEY, G., TURNBULL, A. C. &

GRIFFITHS, K. (1971). The conversion *in vitro* of C_{19}-steroids to oestrogen sulphates by the sheep placenta. *J. Endocr.* **50**, 537–538.

SHUTT, D. A. & COX, R. I. (1972). Steroid and phyto-oestrogen binding to sheep uterine receptors *in vitro*. *J. Endocr.* **52**, 299–310.

SHUTT, D. A. & COX, R. I. (1973). Competitive protein binding assay of oestrone, oestradiol-17β or oestradiol-17α in human or ruminant plasma. *Steroids* (in press).

SHUTT, D. A., COX, R. I. & THORBURN, G. D. (1973). Estrogen changes during late gestation in sheep and goats: relation to parturition. *Endocrinology* (in Press).

THORBURN, G. D. & MATTNER, P. E. (1971). Anastamosis of the utero-ovarian and anterior mammary vein for collection of utero-ovarian venous blood: progesterone secretion rates in cyclic ewes. *J. Endocr.* **50**, 307–320.

THORBURN, G. D., NICOL, D. H., BASSETT, J. M., SHUTT, D. A. & COX, R. I. (1972). Parturition in the goat and sheep: changes in corticosteroids, progesterone, oestrogens and prostaglandin F. *J. Reprod. Fert.* Suppl. **16**, 61–84.

THORBURN, G. D. & SCHNEIDER, W. (1972). The progesterone concentration in the plasma of the goat during the oestrous cycle and pregnancy. *J. Endocr.* **52**, 23–36.

VAN RENSBURG, S. J. (1971). Reproductive physiology and endocrinology of normal and habitually aborting Angora goats. *Onderstepoort J. vet Res.* **38**, 1–62.

HORMONAL INTERACTIONS IN THE
MECHANISM OF PARTURITION

By G. C. LIGGINS

During the 40 years since the discovery of the ovarian hormones, their respective roles in the control of the contractile activity of uterine smooth muscle have been extensively investigated and a number of broad generalisations on their basic actions have become deeply entrenched in the lore of parturitional physiology. Workers in this field have usually felt obliged to explain the results of their experiments in terms of these 'sacred cows', although maintaining spirited controversy by emphasising the particular importance of one or another of the various hormones. In this paper some of the tenets of parturitional physiology will be examined in turn as they apply to studies of the mechanisms controlling the initiation of parturition in sheep. The need for reappraisal has become apparent following the emergence of two new facets of the problem. The first of these is the appreciation of the extent to which the foetus may participate in the endocrine control of parturition; the second is the discovery that prostaglandins probably merit a prominent position in the list of agents influencing uterine smooth muscle at the time when labour starts.

'A waning influence of progesterone on the myometrium is a prerequisite for the start of labour'

The original experiments in pregnant rabbits which showed that ovariectomy caused abortion and that progesterone treatment after ovariectomy prevented abortion (Corner & Allen, 1929) established a strong case for the maintenance of pregnancy being dependent on adequate levels of progesterone. When subsequently it was found that administered progesterone prolonged pregnancy in intact pregnant rabbits (Heckel & Allen, 1937) and that parturition was normally preceded by a fall in the levels of circulating progesterone (Mikhail, Noall & Allen, 1961) the case for progesterone withdrawal as a prerequisite for labour in rabbits was considered proven (Csapo, 1969). When these same tests are applied to pregnant sheep, the criteria for progesterone withdrawal as a prerequisite to labour are met just as they are in the rabbit. Ovariectomy of pregnant sheep before mid-pregnancy causes abortion (Casida & Warwick, 1945) and administration of progesterone maintains pregnancy after ovariectomy. Furthermore, not only may administration of progesterone to intact ewes at term prolong pregnancy (Bengtsson & Schofield, 1963) but also parturition at term is normally preceded by a fall in circulating levels of progesterone (Bassett, Oxborrow, Smith & Thorburn, 1969).

We have made observations of the changes in progesterone levels occurring when labour is initiated under certain experimental conditions which raise some doubts that withdrawal of progesterone is, in fact, a necessary requirement for the operation of the normal mechanisms controlling the initiation of labour.

Our experimental model is the pregnant sheep in which a foetal jugular vein or carotid artery is chronically cannulated and in which continuous infusion of a glucocorticoid (usually dexamethasone) induces premature delivery after a latent period of about 48 h (Liggins, 1969; Liggins, Grieves, Kendall & Knox, 1972; Thorburn et al. 1972). Administering progesterone to the ewe while infusing the foetus with dexamethasone may modify this response in one of three ways. First, in daily doses of 100 mg or less, progesterone does not delay the onset of premature labour or influence the rate of progress of labour. Secondly, in daily doses exceeding 200 mg, progesterone completely inhibits uterine contractions and delivery fails to occur. Thirdly, an intermediate dose of progesterone (150 mg/24 h) does not delay the onset of labour but cervical dystocia occurs and labour is greatly protracted; usually, the membranes rupture and eventually intrauterine infection occurs, leading ultimately to the death of the foetus. Measurements of the plasma concentration of progesterone after its administration to the mother are shown in Fig. 1. It can be seen that parturition occurs without delay even when the amount of injected progesterone is sufficient to fully compensate for diminished placental secretion.

Similar observations were made in normal sheep near term by Bengtsson & Schofield (1963). They noted that progesterone in daily doses of 80 mg usually failed to prevent parturition at term but with doses of 160 mg/24 h some ewes delivered normally at term, some ruptured the membranes near term but did not deliver their lambs and others had prolonged pregnancies. Plasma concentrations of progesterone were not determined but the doses of progesterone they employed exceeded the various estimates of daily placental production rates of 14 mg (Linzell & Heap, 1968), 33 mg (Mattner & Thorburn, 1971) and 12–100 mg (Slotin, Harrison & Heap, 1971). Bengtsson and Schofield concluded from their results that the placental contribution of progesterone was not replaceable by progesterone administered systemically. This implies that the concentration of

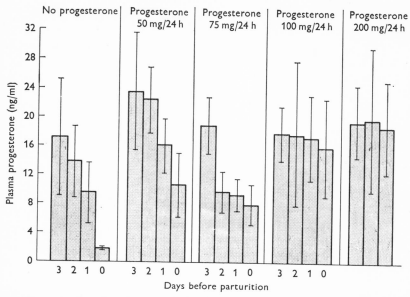

Fig. 1. Concentration of progesterone in jugular plasma of pregnant ewes at 110–130 days of pregnancy during continuous infusion of dexamethasone (1 mg/24 h) into a foetal carotid artery. Progesterone in oil was injected s.c. in the daily doses shown. Values during the progesterone treatment are minimal, samples being obtained immediately before the progesterone injections. Day 0 indicates day of parturition and with the exception of the ewes treated with 200 mg progesterone/24 h, which failed to begin labour, all the ewes delivered on this day. The vertical lines indicate ±s.d.

progesterone in the myometrium falls in those sheep which deliver at term in spite of the fact that enough progesterone was administered to maintain stable circulating levels. Such was not the case in our experiments with induced premature labour since measurements of myometrial progesterone concentration in progesterone-treated animals in labour showed no fall in values compared with normal controls (Liggins *et al.* 1972).

Clearly, in sheep, dosage must be taken into account when drawing conclusions from experiments in which attempts are made to prolong pregnancy by progesterone administration. Unless the concentrations of progesterone in plasma and myometrium are known, it may be difficult to distinguish doses that are within the physiological range from those that are pharmacological.

A major difference exists in the parturitional physiology of rabbits and sheep. In rabbits, functional corpora lutea are necessary for maintenance of pregnancy to term whereas in sheep, pregnancy continues in the second half of gestation after removal of the ovaries. Thus it is possible that there is a fundamental difference in the way that administered progesterone acts in the two species. Indeed, Csapo (1969) has postulated a local action of placental progesterone on the myometrium to explain the absence of effects of large doses of progesterone on the duration of pregnancy in certain species. Although anatomical considerations lead to difficulties in accepting direct transport of progesterone from the trophoblast, where it is synthesized, to the myometrium (Short, 1969), local effects of progesterone on tissues contiguous with the trophoblast are more feasible.

The goat is a species that is of particular interest in providing a possible link between sheep and rabbits. On the one hand, not only is the goat a close relative of the sheep, but also studies of the role of the foetal goat in parturition have shown it to be similar to that of the foetal lamb (Thorburn *et al.* 1972). On the other hand, the goat is analogous to the rabbit in being dependent on ovarian function throughout pregnancy and the major source of progesterone production throughout pregnancy is the corpus luteum (Linzell & Heap, 1968; Thorburn & Schneider, 1972). Thorburn *et al.* (1972) found that daily administration of 20 mg progesterone to goats in which the foetus was infused with Synacthen prevented the expected fall in progesterone levels and delayed parturition until progesterone injections were discontinued. The progesterone treatment maintained the plasma concentration of progesterone within the range found in normal pregnancy. The dose of progesterone used by these authors exceeded the value of 6·1 mg/24 h for the daily production of progesterone observed by Linzell & Heap (1968) in a goat with a single foetus at 124 days of pregnancy and may, therefore, have exerted pharmacological effects. Furthermore, we observed in the sheep that plasma levels of progesterone remained within physiological limits for normal pregnancy even when large doses of progesterone were given (Fig. 1). Thus, the physiological levels of progesterone observed in the goat after treatment with progesterone provide no guarantee that dosage was physiological. Nevertheless, with these reservations, the results suggest that withdrawal of progesterone is a necessary requirement for parturition in the goat. This difference between sheep and goats may be related to the different means by which the foetus controls production of oestrogens and progesterone. In the sheep, it appears that foetal cortisol influences placental production of both oestrogen and progesterone, whereas in the goat, although it apparently increases the production of oestrogen by the placenta, its action on the production of progesterone is ultimately observed in the corpus luteum. The mediator of this effect of foetal cortisol on corpus luteum function is uncertain but there is reason to believe that it may be prostaglandin $F_{2\alpha}$ (Thorburn *et al.* 1972). In the sheep, the maternal placenta is the major site where prostaglandin $F_{2\alpha}$ ($PGF_{2\alpha}$) is synthesised in response to high levels of foetal cortisol (Liggins & Grieves, 1971).

Release of $PGF_{2\alpha}$ from the maternal placenta in sheep may be inhibited by progesterone (see below) and the same may be true of goats. However, an important difference could exist between the two species in relation to the action of progesterone on $PGF_{2\alpha}$ synthesis. In the sheep, the maternal placenta is presumably exposed to a high concentration of progesterone as the latter traverses the maternal tissues from the trophoblast to enter the maternal circulation. In such circumstances, a fall in placental production of progesterone would cause a marked fall in concentration of progesterone in the maternal placenta such that it could be compensated for only by relatively large doses of progesterone. In the goat, however, the maternal placenta is exposed almost exclusively to circulating progesterone and it would make little difference whether the progesterone was of endogenous or exogenous origin. Similar considerations might also apply to the rabbit and other corpus luteum-dependent species.

Table 1. *Effects of maternal progesterone administration on the concentration of $F_{2\alpha}$ ($PGF_{2\alpha}$) in tissues and uterine venous plasma of sheep after continuous infusion of dexamethasone (1 mg/24 h) into the foetus (means $\pm$ S.D.). (The assay was by gas–liquid chromatography.)*

| | Concentration of $PGF_{2\alpha}$ | | |
Treatment group*	Maternal placenta (ng/g)	Myometrium (ng/g)	Uterine vein plasma (ng/ml)
Untreated controls (10)	158 ± 43	130 ± 85	N.D.
Dexamethasone, 48 h infusion (5)	630 ± 152	366 ± 139	24 ± 15·6
Dexamethasone, 48–72 h infusion + progesterone, 200 mg/24 h (5)	N.M.	N.M.	N.D.
Dexamethasone, 48 h infusion + medroxyprogesterone, 600 mg/24 h (5)	501 ± 127	332 ± 87	N.D.

N.D. = not detected; N.M. = not measured
* Numbers of animals in each group are shown in parentheses.

To test the hypothesis that placental progesterone may inhibit release of $PGF_{2\alpha}$ from maternal placenta it is probably necessary to give large doses of progesterone to the ewe to achieve concentrations of progesterone within the maternal cotyledons approximating those normally prevailing as a result of synthesis in adjacent trophoblast. Accordingly, experiments were performed in which the concentrations of $PGF_{2\alpha}$ in maternal placenta, myometrium and uterine venous blood after foetal administration of dexamethasone were compared with similarly-treated animals receiving either progesterone (200 mg/24 h) or medroxyprogesterone acetate (600 mg/24 h) (Liggins, 1973). The results in Table 1 and Fig. 2 show that the concentrations of $PGF_{2\alpha}$ in the tissues of the progesterone-treated and

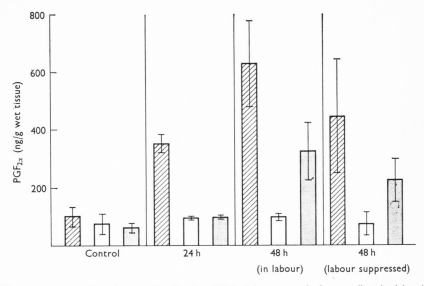

Fig. 2. Concentration of prostaglandin $F_{2\alpha}$ ($PGF_{2\alpha}$) in maternal placenta (hatched bars), foetal placenta (white bars) and myometrium (stippled bars) determined by gas–liquid chromatography (Liggins & Grieves, 1971). All foetuses were continuously infused with 1 mg dexamethasone/24 h for 24 h ($n = 2$), until labour started ($n = 5$) or for 48 h with labour suppressed by progesterone injections ($n = 3$). Tissues were obtained at Caesarean section and were rapidly frozen after separation. In the group in which labour was suppressed by treatment with progesterone (200 mg/24 h) tissues were taken at a time when untreated ewes were in labour. Ewes at 110–130 days of pregnancy. The vertical lines indicate $\pm$s.d. Reproduced with permission from Liggins et al. (1972).

untreated groups of animals were elevated to the same extent by foetal dexamethasone but whereas $PGF_{2\alpha}$ rose sharply in the uterine venous blood of the untreated group, levels in progesterone- and medroxyprogesterone-treated ewes remained low. The significance of this observation is not certain but it seems likely that it indicates that progesterone caused a marked reduction in the rate of synthesis of $PGF_{2\alpha}$ in uterine and placental tissues. The rate of turnover of prostaglandins in tissues is thought to be high (Pickles, Hall, Best & Smith, 1965) and the concentration of prostaglandin in a tissue may, therefore, give little information about the rate of synthesis. On the other hand, the concentration of prostaglandin in blood leaving the organ where it was formed probably gives some indication of production rate even though it ignores the likelihood of degradation within the tissue.

Further evidence favouring inhibition by progesterone of $PGF_{2\alpha}$ synthesis was found in two experiments in which the rise in concentration of $PGF_{2\alpha}$ in uterine venous blood resulting from treatment with stilboestrol (see next section) was prevented by simultaneous treatment with progesterone (Fig. 5).

If the assumption is accepted that in corpus luteum-dependent species labour occurs only when the influence of progesterone is withdrawn, eutherian mammals can be divided into three classes:

1. Corpus luteum-dependent species in which labour must be preceded by a fall in circulating levels of progesterone. These species may respond with prolongation of pregnancy after doses of progesterone that are within physiological limits (e.g. rabbits, goats).

2. Placenta-dependent species in which plasma progesterone levels are relatively low throughout pregnancy. Prolonged pregnancy in such mammals may occur only in response to pharmacological doses of progesterone (e.g. sheep, cattle). It is possibly relevant that these species have syndesmochorial placentae in which foetal and maternal tissues are in direct contact, permitting local effects of placental progesterone on maternal tissues.

3. Placenta-dependent species in which plasma progesterone levels are high throughout pregnancy due to elevated levels of binding proteins in the plasma. Pregnancy is unaffected by high doses of progesterone (e.g. man, guinea-pigs). In these species maternal tissues in the placenta are separated by blood from foetal tissues and local effects of progesterone are less likely.

In summary, this section has considered the possibility that a major site of action of progesterone is in the maternal placenta (decidua). This site may be susceptible to a local influence of placental progesterone which could account for some of the variation between species in their response to exogenous progesterone.

'Labour occurs when the ratio of uterine volume to myometrial progesterone concentration increases beyond a critical value'

According to this hypothesis, the ratio $V:P$ (where V = uterine volume and P = myometrial progesterone concentration) is the key regulatory mechanism in the onset of parturition. Bedford, Challis, Harrison & Heap (1972) point out that in view of the gradual nature of the changes in V and P in many species it is unlikely that the $V:P$ ratio alone can account for the precise events associated with the onset of parturition. For example, it can be calculated from available data on amniotic fluid volume and foetal and placental weights that the volume of the human uterus normally increases by no more than 2 per cent per week in the last month of pregnancy. On the other hand, although the association of premature labour and polyhydramnios is often quoted in support of the $V:P$ hypothesis, the fact remains that the human uterus may accommodate 10

litres or more of amniotic fluid despite normal levels of circulating progesterone before labour finally starts. Similarly, the association of multiple
pregnancy with premature labour in women could equally well be attributed to the effects of some component of foetal mass as shown in mice
by McLaren (1967) rather than to uterine distension. Furthermore, distension of the uterus of the non-pregnant guinea-pig causes release of
$PGF_{2\alpha}$ (Poyser, Horton, Thompson & Los, 1971) and the possibility cannot
be excluded that unduly rapid or excessive distension of the pregnant uterus
has a similar effect.

In sheep we have observed extreme polyhydramnios (in excess of 11
litres) together with foetal hydrops after foetal decapitation, yet the
pregnancies were prolonged (unpublished observations). Multiple pregnancy has a negligible effect on duration of pregnancy in sheep (Terrill
& Hazel, 1947) but progesterone levels are higher than in single pregnancies
(Bassett et al. 1969) and $V:P$ may be the same as in single pregnancies.
The important role played by the foetal endocrine system in controlling
parturition in sheep leaves little scope for changes in uterine volume to
make a significant contribution to the mechanism. Likewise, altered
progesterone levels appear to play only a minor part.

'Quiescence of the pregnant uterus and unresponsiveness to oxytocin are due to the progesterone-dominance of the myometrium'

According to Marshall & Csapo (1961) the progesterone-dominated
rabbit myometrium has a raised threshold to stimuli and the propagation
of depolarisation is impaired. The experimental evidence for these con-
tentions has been reviewed by Carsten (1968) and they will not be considered further here.

We have studied uterine activity and the response to oxytocin in pregnant
sheep at 110–130 days of gestation in which the uterus was progesterone-
dominated as judged by measurements of progesterone concentration in
both peripheral blood and uterine vein blood (Liggins, 1973). The effects
either of a single subcutaneous injection into the ewe of 20 mg stilboestrol
(in oil) or of a continuous intra-aortic infusion of $PGF_{2\alpha}$ on spontaneous
and oxytocin-induced uterine activity were studied. Intra-amniotic
pressure was recorded continuously by means of an intra-amniotic balloon
catheter connected to a Statham pressure transducer and a potentiometric
recorder. The threshold dose of oxytocin was determined at intervals of
12 h by rapidly injecting oxytocin intravenously in progressively increasing
doses starting with 10 mu. until a response was obtained. At the same time,
blood samples were taken both from the uterine vein for assay of proges-

terone and $PGF_{2\alpha}$ and from the jugular vein for assay of oestradiol-17β and progesterone. Finally, the ewes treated with stilboestrol were killed and placental and myometrial tissues were taken for assay of $PGF_{2\alpha}$. The results of the experiments are shown in Figs. 3–7. It is apparent that

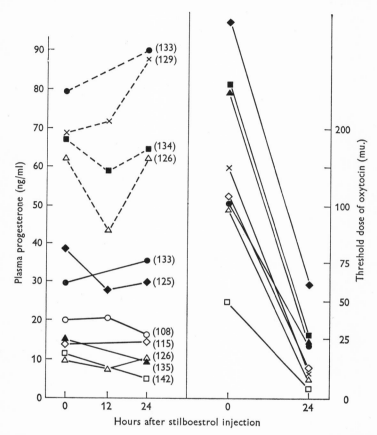

Fig. 3. Changes in oxytocin sensitivity and in the concentration of progesterone in maternal jugular (———) and uterine (– – – –) venous plasma after single s.c. injections of 20 mg stilboestrol in oil into nine ewes. On the left the progesterone concentrations are shown; on the right the threshold doses of oxytocin in eight of the ewes. The threshold dose of oxytocin is the smallest amount, given as a single i.v. injection, causing a definite response on a continuous recording of intra-amniotic pressure. Gestation length in days is given in parentheses. Each sheep is represented by the same symbol in both parts of the figure.

treatment with oestrogen or $PGF_{2\alpha}$ caused not only a sharp fall in oxytocin threshold but also the onset of regular uterine contractions without a concurrent fall in either the levels of progesterone in peripheral blood or, in the case of oestrogen treatment, in the production of progesterone by the placenta. The concentration of $PGF_{2\alpha}$ in uterine venous blood and in tissues of the maternal placenta and myometrium rose sharply after the injection of oestrogen (Figs. 4 and 7). The uterine responses to $PGF_{2\alpha}$ and to stilboestrol are similar and since synthesis of $PGF_{2\alpha}$ is stimulated by oestrogen, the effects of oestrogen on uterine smooth muscle may be mediated by $PGF_{2\alpha}$.

In the above experiments, pregnancy was interrupted by Caesarean section after 24–48 h to obtain tissues for assay. Hindson, Schofield &

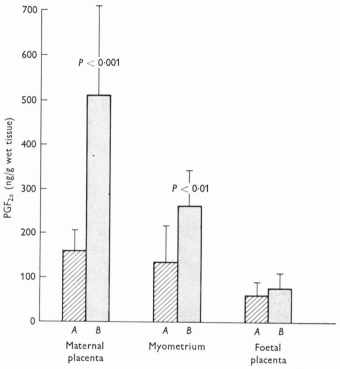

Fig. 4. Concentration of prostaglandin $F_{2\alpha}$ ($PGF_{2\alpha}$) in maternal placenta, myometrium and foetal placenta 24 h after an injection of 20 mg stilboestrol in oil into ewes at 110–130 days of gestation. See legend to Fig. 2 for method of assay. A = controls; B = ewes treated with stilboestrol. The vertical lines indicate the s.D. $n = 8$.

Turner (1967) gave single injections of 20 mg stilboestrol to ewes at 132–142 days of gestation and noted that uterine contractions appeared exactly 24 h later. Contractions subsequently ceased in the ewe treated at 132 days but in the remainder, all beyond 135 days of pregnancy, contractions continued and delivery occurred. In some of the ewes, cervical dystocia was troublesome and Caesarean section or manual dilatation of the cervix became necessary to achieve delivery. A similar type of cervical dystocia

is observed when labour starts in the presence of raised levels of progesterone (Liggins *et al.* 1972). Although progesterone assays were not
performed in the experiments of Hindson and coworkers it is possible that
progesterone levels were maintained during labour in their ewes as was
observed in our experiments in which stilboestrol was given at 110–130
days of gestation. This suggests that the importance of any direct actions
of progesterone on uterine smooth muscle lies less in a relationship to the

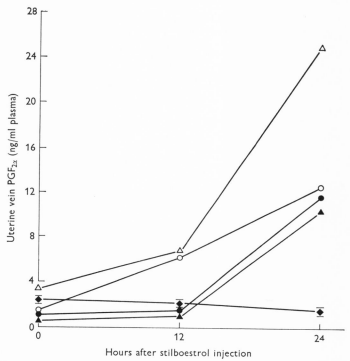

Fig. 5. Concentration of prostaglandin $F_{2\alpha}$ ($PGF_{2\alpha}$) in uterine venous blood of ewes after
an injection of 20 mg stilboestrol in oil. In three experiments (♦) 200 mg progesterone/24 h
was infused i.v. from the time when stilboestrol was given. The horizontal bars indicate
±S.D. $PGF_{2\alpha}$ was determined by the radioimmunoassay of Grieves & Kendall (1972).

initiation of labour than in a relationship to myometrial function during
labour.

If the prime action of progesterone in the ewe is related to inhibition of
synthesis of $PGF_{2\alpha}$ it should follow that effects on the myometrium usually
attributed to a direct action of progesterone on smooth muscle will be
reversed by $PGF_{2\alpha}$ and that progesterone will not block changes in
myometrial contractility induced by $PGF_{2\alpha}$. According to Csapo (1969)
the sensitivity of the myometrium to oxytocin is determined by the degree

of 'progesterone dominance' and measurement of oxytocin threshold is considered to be the most reliable index of progesterone influence on the myometrium (Schofield, 1962). However, we found that the oxytocin threshold diminished sharply during infusions of $PGF_{2\alpha}$ in ewes although

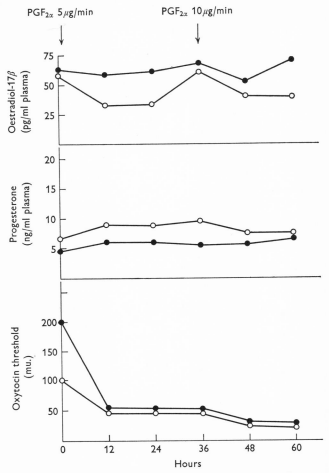

Fig. 6. Changes in oxytocin sensitivity and in the concentration of progesterone and oestradiol-17β in maternal jugular venous plasma during continuous intra-aortic infusion of prostaglandin $F_{2\alpha}$ ($PGF_{2\alpha}$) at a rate of 5–10 μg/min into two ewes. Ewes at 118–127 days of pregnancy.

no significant change in circulating levels of progesterone occurred. Moreover, continuous administration of progesterone at a rate of 200 mg/24 h failed to restore the lowered threshold to its former level (Figs. 8 and 9). These observations are consistent with the proposal that oxytocin sensitivity is determined by $PGF_{2\alpha}$ rather than by progesterone and that the apparent direct effects of progesterone are mediated by $PGF_{2\alpha}$.

Under the conditions of our experiments, not only was increased myometrial sensitivity to oxytocin independent of the production of pro-

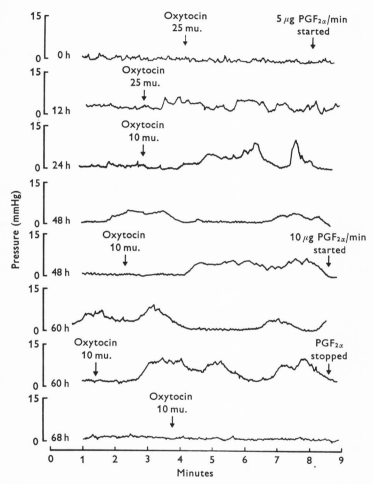

Fig. 7. Recordings of intra-amniotic pressure during a continuous intra-aortic infusion of prostaglandin $F_{2\alpha}$ ($PGF_{2\alpha}$) in a ewe at 127 days of pregnancy. The oxytocin responses are to threshold doses injected i.v. Spontaneous activity approximating that of normal labour appeared after 24 h.

gesterone but also labour-like contractions were observed in the presence of maintained levels of progesterone. The same changes in myometrial behaviour occur near term without a concomitant fall in progesterone levels in women (Kumar, Ward & Barnes, 1964; Llauro, Runnebaum &

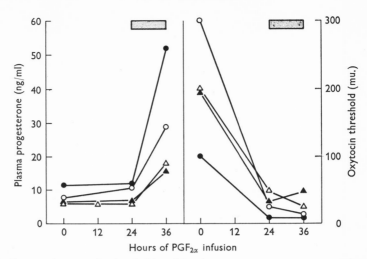

Fig. 8. Changes in oxytocin threshold and in plasma progesterone concentration in four pregnant ewes in which labour was induced by intra-aortic infusion of prostaglandin $F_{2\alpha}$ (PGF$_{2\alpha}$) at a rate of 10 μg/min. During the periods indicated by the horizontal bars, 200 mg progesterone/24 h was infused into a jugular vein.

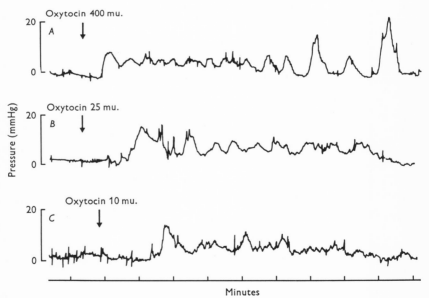

Fig. 9. Intra-amniotic pressure recording during continuous intra-aortic infusion of prostaglandin $F_{2\alpha}$ (PGF$_{2\alpha}$) at a rate of 10 μg/min, illustrating the lack of effect of progesterone on oxytocin threshold in sheep. A = before treatment; B = after 24 h infusion of PGF$_{2\alpha}$; C = after 48 h infusion of PGF$_{2\alpha}$ and 12 h continuous i.v. infusion of progesterone at a rate of 200 mg/24 h.

Zander, 1968; Ances, Hisley & Haskins, 1971) and in guinea-pigs (Schofield, 1964; Heap & Deanesly, 1967; Challis, Heap & Illingworth, 1971). In species like the sheep in which the myometrial changes normally occur at a time of diminishing progesterone levels the possibility must be considered that other factors such as the myometrial concentration of prostaglandins are contributing to the various alterations in myometrial function usually ascribed to withdrawal of progesterone and that the concept of 'progesterone dominance' needs modification.

> *'Spontaneous activity of the pregnant uterus and increased responsiveness to oxytocin are due to the oestrogen-dominance of the myometrium.'*

The chronic effects of oestrogens on the pregnant uterus *in vivo* are well documented. There seems no doubt that oestrogens are bound specifically to a cytoplasmic receptor and transported into the nucleus where they promote protein synthesis. The acute effects of oestrogen, on the other hand, are unclear. Despite the fact that the action of oestrogen on the pregnant rat and rabbit uterus *in vitro* is inhibitory (Saldivar & Melton, 1966) it is commonly asserted that oestrogen has oxytocic properties *in vivo* (Pinto, Leon, Mazzocco & Scasserra, 1967; Bedford *et al.* 1972) or at least causes increased responsiveness to oxytocin (Jarvinen, Luukkainen & Vaisto, 1965).

The response of the uterus of the pregnant ewe to oestrogen in our experiments and those of Hindson *et al.* (1967) appears to support the idea that oestrogen acts directly on the myometrium to stimulate contractions and to increase responsiveness to oxytocin. However, as already discussed, we found evidence suggesting that the action of oestrogen is primarily on the synthesis of $PGF_{2\alpha}$ and that the apparent myometrial effects of oestrogen are, in fact, responses to $PGF_{2\alpha}$. If the only stimulus to prostaglandin synthesis were oestrogens then there might seem little point in differentiating between the effects of oestrogen and those of $PGF_{2\alpha}$ but it is possible that prostaglandin synthesis can be influenced in other ways. Thus, the term 'oestrogen-dominance' could be misleading.

> *'Oxytocin must be released in increased quantities if it is to contribute to parturitional mechanisms.'*

It has been calculated by Fitzpatrick (1966) that a secretion rate of oxytocin in women of 10 mu./min corresponds to a concentration of about 20 μu./ml in internal jugular venous blood. In peripheral blood the concentration is much lower due to the effects of dilution and degradation and it is below the limits of detection by present assay methods (Chard, Hudson, Edwards & Boyd, 1971). The rapid rise of oxytocin observed in several species including sheep (Fitzpatrick & Walmsley, 1965) during the expulsive phase of labour is generally accepted as having a function in stimulating uterine activity in the final stage of labour but the absence of a measurable increase in levels at the onset of labour is likewise taken as evidence against oxytocin having an important role in the initiation of parturition (Caldeyro-Barcia, Melander & Coch, 1971). In rabbits the hypothesis that oxytocin

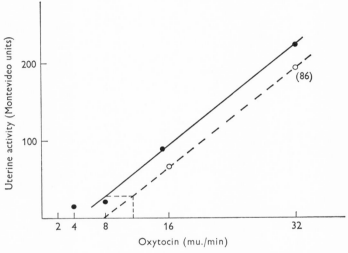

Fig. 10. Oxytocin dose–response curves in a patient with diabetes insipidus near term. The response was determined before (●) and during (○) an i.v. infusion of ethanol. The figure in parentheses is blood alcohol concentration (mg/100 ml). The second regression line is displaced to the right by a distance equivalent to an infusion rate of 3 mu./min. Reproduced with permission from Mantell & Liggins (1970).

initiates labour (Fuchs, 1964) rests on evidence of an abrupt and relatively massive release of oxytocin from the posterior pituitary gland at a time when little uterine activity has developed.

In studying the effects of ethanol on uterine activity induced in women near term by infusions of oxytocin, Mantell & Liggins (1970) noted that the effect of ethanol was to shift the oxytocin dose–response curve to the right. They argued that since ethanol is thought to inhibit release of oxytocin, the displacement of the curve is the result of a fall in endogenous oxytocin levels, although in some women loss of parallelism suggests an additional direct non-competitive inhibitory effect of ethanol on the myometrium. When parallelism was present, they considered that the extent of displacement of the dose–response curve is a measure of endogenous oxytocin. It was calculated that circulating oxytocin levels in their patients (who were not in labour) corresponded to a secretion rate of 2–8 mu./min. Of particular interest was a patient suffering from diabetes insipidus in whom it was estimated that the rate of secretion of oxytocin was approximately 3 mu./min (Fig. 10). If 5 mu./min is assumed as an approximate value for the secretion rate of oxytocin in normal women in late pregnancy and Fitzpatrick's calculation is applied to it, the concentration of oxytocin in internal jugular blood would be 10 μu./ml. Cerebral blood flow represents about 10 per cent of cardiac output and the effects of dilution together with the short half-life of oxytocin will reduce its concentration to levels near to the limits of assay of 1μu./ml (Chard et al. 1971). Furthermore, Theobald, Robards & Suter, (1969) found that the sensitivity to oxytocin infused at rates of only 0·5–1 mu./min increased greatly before the onset of labour. Thus, in women, concentrations of circulating oxytocin below the present limits of detection may well have significant effects on uterine contractility at the onset of labour. Nevertheless, the cause of labour according to this argument lies with the factors determining oxytocin sensitivity rather than with oxytocin itself.

It is not known whether oxytocin is present in the circulation of ewes near term. Our experiments described above in which increasing sensitivity to oxytocin was associated with administration of oestrogen or $PGF_{2\alpha}$ suggest the possibility that the evolution of uterine contractions during treatment with these agents might be a consequence of the development of a uterine response to a constant level of oxytocin. This receives some support from the observation that uterine contractions appeared only after 24 h or more of infusion with $PGF_{2\alpha}$, making it unlikely that the response was a direct oxytocic action of $PGF_{2\alpha}$. At the time when uterine contractions first started the only hormonal change found was in the myometrial sensitivity to oxytocin; oestrogen and progesterone levels were unaltered.

The fact that hypophysectomised ewes deliver normally at term (Denamur & Martinet, 1961) does not negate the possibility that oxytocin plays a part in the initiation of parturition. Women with diabetes insipidus may continue to secrete oxytocin (Mantell & Liggins, 1970) although their ability to release the relatively large amounts necessary to induce the 'let-down' of milk is impaired. The hypothalamus of hypophysectomised ewes could also be capable of maintaining basal secretion of oxytocin as it apparently does for the antidiuretic hormone.

CONCLUSION

The various observations of the endocrine changes associated with the initiation of parturition in sheep would be interpreted, according to classical teachings, as follows:

> Elevated levels of foetal cortisol resulting from an increase in the rate of secretion from the foetal adrenal cortex act on the placenta to cause a fall in the production of progesterone and an increase in the secretion of unconjugated oestrogen. Thus the state of progesterone dominance is altered to one of oestrogen dominance. Consequently, the inherent tendency of the uterus to contract spontaneously is permitted to emerge and the myometrium is further stimulated by oestrogen to a point where labour starts. When the birth canal becomes distended, a neural reflex stimulates the release of oxytocin which augments uterine contractions and leads to delivery.

An alternative explanation possibly worthy of further consideration is as follows:

> Foetal cortisol by unknown means stimulates release of unconjugated oestrogen from the placenta and causes a fall in secretion of progesterone. The effect of oestrogen in the presence of progesterone is to stimulate the potential for synthesis of $PGF_{2\alpha}$ in the maternal placenta while the local action of progesterone is to block the release of $PGF_{2\alpha}$. The net effect of the changes in oestrogen and progesterone production preceding parturition is release from the maternal placenta (decidua) of large amounts of $PGF_{2\alpha}$ which reach the myometrium by an unknown, but direct, route. The action of $PGF_{2\alpha}$ on the myometrium is to increase its sensitivity to circulating oxytocin and this action is not dependent on any changes in levels of progesterone or unconjugated oestrogens. $PGF_{2\alpha}$ may also have direct oxytocic effects at physiological concentrations. It will be noted that this hypothesis calls for no direct action of either oestrogen or progesterone on the uterine smooth muscle cell. On the contrary, the acute actions of both oestrogen and progesterone are postulated to occur in the basal tissues ('decidua') of the maternal placenta where they control release of $PGF_{2\alpha}$. Labour starts as a consequence of these changes in myometrial $PGF_{2\alpha}$. Uterine contractions become more effective in dilating the cervix when myometrial concentrations of progesterone fall, as they normally do, due to the effects of foetal cortisol on placental progesterone production. However, the major action of progesterone is in inhibiting synthesis of $PGF_{2\alpha}$, particularly by a local action on the maternal placenta. Finally, labour is augmented by reflex release of oxytocin (Fig. 11).

In effect, it is suggested that in the sheep the major *acute* actions of

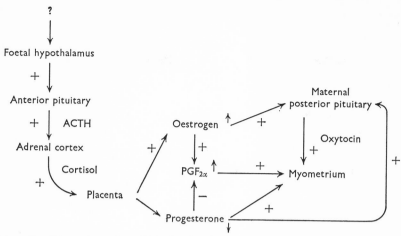

Fig. 11. Possible pathway by which the foetal endocrine system influences myometrial contractility at the initiation of parturition. In the present communication the hypothesis is proposed that the action of oestrogen and progesterone are mediated by their respective stimulatory and inhibitory effects on the synthesis of prostaglandin $F_{2\alpha}$ ($PGF_{2\alpha}$). Thus the direct pathway shown for a direct action of progesterone on the myometrium is considered as a minor one under physiological conditions although it may become dominant as a result of pharmacological levels of progesterone.

Plus and minus signs represent positive and negative effects. In the case of oestrogen, progesterone and $PGF_{2\alpha}$, the plus sign represents a positive effect resulting from a change in the level of the hormone in the direction indicated by the small vertical arrow placed at the end of the word. ACTH = adrenocorticotrophin.

oestrogen and progesterone are on the biosynthesis and release of $PGF_{2\alpha}$ rather than directly on the contractile mechanisms of the smooth muscle cells.

ACKNOWLEDGEMENTS

I am grateful to my collaborators, Miss Susan Grieves, Miss June Kendall and Mr B. S. Knox for biochemical work and to Mr A. Mekkelholt and Mr. H. Wattam for their care of the animals. The work was supported by generous grants from The Wellcome Trust. Prostaglandins were a gift from Dr J. E. Pike, Upjohn Company, Kalamazoo, Michigan, USA.

REFERENCES

ANCES, I. G., HISLEY, J. C. & HASKINS, A. L. (1971). Studies on the level of blood progesterone throughout the course of labor. *Am. J. Obstet. Gynec.* **109**, 36–40.

BASSETT, J. M., OXBORROW, T. J., SMITH, I. D. & THORBURN, G. D. (1969). The concentration of progesterone in the peripheral plasma of the pregnant ewe. *J. Endocr.* **45**, 449–457.

BEDFORD, C. A., CHALLIS, J. R. G., HARRISON, F. A. & HEAP, R. B. (1972). The role of oestrogen and progesterone in the onset of parturition in various species. *J. Reprod. Fert.* Suppl. **16**, 1.

BENGTSSON, L. PH. & SCHOFIELD, B. M. (1963). Progesterone and the accomplishment of parturition in sheep. *J. Reprod. Fert.* **5**, 423–431.

CALDEYRO-BARCIA, R., MELANDER, S. & COCH, J. A. (1971). Neurohypophyseal hormones. In *Endocrinology of Pregnancy*, eds. F. Fuchs & A. Klopper, pp. 235–285. New York: Harper and Row.

CARSTEN, M. E. (1968). Regulation of myometrial composition, growth and activity. In *Biology of Gestation*, ed. N. S. Assali, vol. 1, pp. 356–425. New York: Academic Press.

CASIDA, L. E. & WARWICK, E. J. (1945). The necessity of the corpus luteum for maintenance of pregnancy in the ewe. *J. Anim. Sci.* **4**, 34–36.

CHALLIS, J. R. G., HEAP, R. B. & ILLINGWORTH, D. V. (1971). Concentrations of oestrogen and progesterone in the plasma of non-pregnant, pregnant and lactating guinea-pigs. *J. Endocr.* **51**, 333–345.

CHARD, T., HUDSON, C. N., EDWARDS, C. R. W. & BOYD, N. R. H. (1971). Release of oxytocin and vasopressin by the human foetus during labour. *Nature, Lond.* **234**, 352–354.

CORNER, G. W. & ALLEN, W. M. (1929). Physiology of the corpus luteum. *Am. J. Physiol.* **88**, 326–339.

CSAPO, A. I. (1969). The four direct regulatory factors of myometrial function. In *Ciba Foundation Study Group No. 34*, eds. G. E. W. Wolstenholme & J. Knight, pp. 13–55. London: J. & A. Churchill.

DENAMUR, R. & MARTINET, J. (1961). Effets de l'hypophysectomie et de la section de la tige pituitaire sur la gestation de la brebis. *Annls. Endocr.* **21**, 755–759.

FITZPATRICK, R. J. (1966). The posterior pituitary gland and the female reproductive tract. In *The Pituitary Gland*, eds. G. W. Harris & B. T. Donovan, p. 484. London: Butterworths.

FITZPATRICK, R. J. & WALMSLEY, C. F. (1965). The release of oxytocin during parturition. In *Advances in Oxytocin Research*, ed. J. H. M. Pinkerton, pp. 57–71. London: Pergamon Press.

FUCHS, A.-R. (1964). Oxytocin and the onset of labour in rabbits. *J. Endocr.* **30**, 217–224.

GRIEVES, S. A. & KENDALL, J. Z. (1972). Radioimmunoassay of prostaglandin $F_{2\alpha}$. *N.Z. Med. J.* (in Press).

HEAP, R. B. & DEANESLY, R. (1967). The increase in plasma progesterone levels in the pregnant guinea-pig and its possible significance. *J. Reprod. Fert.* **14**, 339–341.

HECKEL, G. P. & ALLEN, W. M. (1937). Prolongation of pregnancy in the rabbit by injection of progesterone. *Am. J. Physiol.* **119**, 330.

HINDSON, J. C., SCHOFIELD, B. M. & TURNER, C. B. (1967). The effect of a single dose of stilboestrol on cervical dilatation in pregnant sheep. *Res. vet. Sci.* **8**, 353–360.

JARVINEN, P. A., LUUKKAINEN, T. & VAISTO, L. (1965). The effect of oestrogen treatment on myometrial activity in late pregnancy. *Acta obstet. gynec. scand.* **44**, 258–264.

KUMAR, D., WARD, E. F. & BARNES, A. C. (1964). Serial plasma progesterone levels and onset of labour. *Am. J. Obstet. Gynec.* **90**, 1360–1361.

LIGGINS, G. C. (1969). Premature delivery of foetal lambs infused with glucocorticoids. *J. Endocr.* **45**, 515–523.

LIGGINS, G. C. (1973). The physiological mechanisms controlling the initiation of ovine parturition. *Recent Prog. Horm. Res.* **29**, 110–149.

LIGGINS, G. C. & GRIEVES, S. A. (1971). Possible role for prostaglandin $F_{2\alpha}$ in parturition in sheep. *Nature, Lond.* **232**, 629–631.

LIGGINS, G. C., GRIEVES, S. A., KENDALL, J. Z. & KNOX, B. S. (1972). The physiological roles of progesterone, oestradiol-17β and prostaglandin $F_{2\alpha}$ in the control of ovine parturition. *J. Reprod. Fert.* Suppl. **16**, 85–103.

LINZELL, J. L. & HEAP, R. B. (1968). A comparison of progesterone metabolism in the pregnant sheep and goat: sources of production and an estimation of uptake of some target organs. *J. Endocr.* **41**, 433–438.

LLAURO, J. L., RUNNEBAUM, B. & ZANDER, J. (1968). Progesterone in human peripheral blood before, during and after labor. *Am. J. Obstet. Gynec.* **101**, 867–873.

McLAREN, A. (1967). Effect of foetal mass on gestation period in mice. *J. Reprod. Fert.* **13**, 349–351.

MANTELL, C. D. & LIGGINS, G. C. (1970). The effect of ethanol on the myometrial response to oxytocin in women at term. *J. Obstet. Gynaec. Br. Commonw.* **77**, 976–981.

MARSHALL, J. M. & CSAPO, A. I. (1961). Hormonal and ionic influences on the membrane activity of uterine smooth muscle cells. *Endocrinology*, **68**, 1026–1035.

MATTNER, P. E. & THORBURN, G. D. (1971). Progesterone in utero-ovarian venous plasma during pregnancy in ewes. *J. Reprod. Fert.* **24**, 140–141.

MIKHAIL, G., NOALL, M. W. & ALLEN, W. M. (1961). Progesterone levels in the rabbit ovarian blood throughout pregnancy. *Endocrinology*, **69**, 504–509.

PICKLES, V. R., HALL, W. J., BEST, F. A. & SMITH, G. N. (1965). Prostaglandins in endometrium and menstrual fluid from normal and dysmenorrhoeic subjects. *J. Obstet. Gynaec. Br. Commonw.* **72**, 185–192.

PINTO, R. M., LEON, C., MAZZOCCO, N. & SCASSERRA, V. (1967). Action of estradiol-17β at term and at onset of labor. *Am. J. Obstet. Gynec.* **98**, 540–546.

POYSER, N. L., HORTON, E. W., THOMPSON, C. J. & LOS, M. (1971). Identification of prostaglandin $F_{2\alpha}$ released by distension of guinea-pig uterus *in vitro*. *Nature, Lond.* **230**, 526–528.

SALDIVAR, J. T. & MELTON, C. E. (1966). Effects *in vivo* and *in vitro* of sex steroids on rat myometrium. *Am. J. Physiol.* **211**, 835–843.

SCHOFIELD, B. M. (1962). The effect of oestrogen on pregnancy in the rabbit. *J. Endocr.* **25**, 95–100.

SCHOFIELD, B. M. (1964). Myometrial activity in the pregnant guinea pig. *J. Endocr.* **30**, 347–354.

SHORT, R. V. (1969). The foetal role in the initiation of parturition in the ewe. In *Ciba Symposium on Foetal Autonomy*, eds. G. E. W. Wolstenholme & M. O'Connor, p. 234. London: J. & A. Churchill.

SLOTIN, C. A., HARRISON, F. A. & HEAP, R. B. (1971). Kinetics of progesterone metabolism in the pregnant sheep. *J. Endocr.* **49**, xxx.

TERRILL, C. E. & HAZEL, L. N. (1947). Length of gestation in range sheep. *Am. J. vet. Res.* **8**, 66.

THEOBALD, G. W., ROBARDS, M. F. & SUTER, P. E. N. (1969). Changes in myometrial sensitivity to oxytocin in man during the last six weeks of pregnancy. *J. Obstet. Gynaec. Br. Commonw.* **76**, 385–393.

THORBURN, G. D., NICOL, D. H., BASSETT, J. M., SHUTT, D. A. & COX, R. I. (1972). Parturition in the goat and sheep: changes in corticosteroids, progesterone, oestrogens and prostaglandin F. *J. Reprod. Fert.* Suppl. **16**, 61–84.

THORBURN, G. D. & SCHNEIDER, W. (1972). The progesterone concentration in the plasma of the goat during the oestrous cycle and pregnancy. *J. Endocr.* **52**, 23–36.

COMPARATIVE ASPECTS OF FACTORS INVOLVED IN THE ONSET OF LABOUR IN OVINE AND HUMAN PREGNANCY

By ANNE B. M. ANDERSON* AND
A. C. TURNBULL*

INTRODUCTION

Many factors affect the contractility of the myometrium in pregnancy and there is increasing evidence that the onset of labour is the result of several interacting influences without any one factor predominating. The theories proposed for this complex regulatory mechanism include endocrine, neurological, mechanical and chemical factors. We do not propose to deal with the last three of these in this paper but rather to discuss the endocrine control of gestational length and to describe our contribution to further understanding of this control mechanism in both ovine and human pregnancy.

In sheep, a great deal of information on the endocrine events associated with parturition has accrued in the past few years, concerned with the effects of foetal corticosteroids on the plasma and tissue levels of other steroids, such as progesterone and oestrogens, and of prostaglandins. Although there is much less information on possible endocrine factors involved in human parturition, and extrapolation of results from sheep to man is difficult, many analogies can be drawn. In this paper we will discuss some of the evidence for the role of corticosteroids, oestrogens, progesterone and prostaglandins in the onset of labour in these two species.

CORTICOSTEROIDS

In sheep

Studies on the metabolism of steroids by the foetal adrenal glands in this species (Anderson, Pierrepoint, Griffiths & Turnbull, 1972) were prompted by evidence that the sheep foetus plays a major role in controlling the initiation of parturition, and that foetal adrenal glucocorticoid synthesis under the control of the foetal pituitary, is the central feature of this

* Address as from 1 October 1973: Maternity Department, The John Radcliffe Hospital, Headington, Oxford, OX3 9DU.

regulating mechanism (Liggins, Holm & Kennedy, 1966; Liggins, 1968, 1969).

The biosynthetic pathways for corticosteroid formation were studied in both adult and foetal sheep adrenal tissue by in-vitro incubation techniques, using isotopically-labelled pregnenolone and progesterone as substrates. The findings suggested that there may be a difference between the steroid biosynthetic pathways of foetal and adult sheep adrenal glands. The foetus appears to have the ability to use both pregnenolone and progesterone for the synthesis of cortisol in the adrenal, in contrast to the adult in which cortisol formation mainly takes place by a route independent of progesterone.

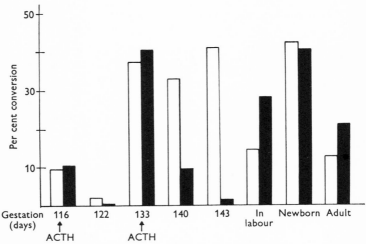

Fig. 1. Percentage conversion of isotopically-labelled pregnenolone to 11-deoxycortisol (white columns) and to cortisol (black columns) after incubation *in vitro* for 2 h of foetal sheep adrenal tissue at 116, 122, 133, 140, 143 days of gestation and on 143 days when the ewe was in early labour. Adrenal tissue from a newborn lamb and from an adult was also studied. At 116 and at 133 days of gestation, synthetic adrenocorticotrophic hormone (ACTH) had been infused into the foetus *in utero* for 72 and 76 h respectively.

The development of steroid-metabolising enzymes in foetal sheep adrenal tissue was investigated *in vitro*, mainly by studying at different stages of gestation, including labour, the ability of foetal adrenal tissue to metabolise radioactive pregnenolone and progesterone to corticosteroids. The effect on this metabolism of synthetic adrenocorticotrophic hormone (ACTH; Synacthen) infused into the foetus *in utero* to induce premature labour was also studied. The conversion of the substrate pregnenolone to 11-deoxycortisol and cortisol after a 2 h incubation period is shown in histogram form in Fig. 1. At 122 days of gestation and in 'full-term' foetal sheep adrenal glands (140 and 143 days) there was a marked inactivity of the 11β-hydroxylating enzyme system for 11-deoxycortisol. At 143 days, for example, 41 per cent of the incubated pregnenolone was converted to 11-deoxycortisol but only 1·3 per cent of the pregnenolone was metabolised to cortisol. In contrast, there was notable 11β-hydroxylase activity in Synacthen-stimulated glands, in the adrenals removed from the foetus during the course of spontaneous labour, and in those from the newborn lamb. In these glands, the transformation of pregnenolone to cortisol was at least as great, if not greater, than to its 11-deoxy precursor. These results are in keeping with the hypothesis that the 11β-hydroxylase activity in the foetal sheep adrenal may be rapidly activated, possibly by foetal ACTH, in the final few days of pregnancy, thus increasing the synthesis of cortisol which in turn may initiate labour. Supportive evidence is provided from experiments *in vivo* where 11-deoxycortisol infused into single lambs *in utero* for 7–8 days was not followed by premature parturition in four out of five cases, whereas similar doses of cortisol infused into the foetus induced premature delivery within 3 days in every case.

In human pregnancy

Although the involvement of foetal corticosteroids in ovine parturition is now well established, evidence for foetal adrenal participation in human parturition is, of necessity, circumstantial and has been obtained for the most part from studies on congenital anomalies.

Anencephaly and prolonged pregnancy

This foetal malformation, mainly of the central nervous system, is almost invariably associated with hypoplasia of the adrenal cortex, probably as a result of failure of functional development of the hypothalamus. The first report of an association between anencephaly and prolongation of gestation came from Rea in 1898, since when several published series have confirmed this observation (Malpas, 1933; Comerford, 1965; Milic & Adamsons, 1969; Anderson, Laurence & Turnbull, 1969). In the absence of hydramnios, the incidence of postmaturity (continuation of pregnancy beyond the 42nd week) can be as high as 60 per cent in association with foetal anencephaly. Furthermore, the extent of the prolongation of pregnancy can be correlated with the degree of hypoplasia of the foetal adrenal glands (Anderson *et al.* 1969) and with increasing difficulty in the induction of labour by means of amniotomy and intravenous oxytocin (Comerford, 1965; Anderson *et al.* 1969). However, despite the marked hypoplasia of the adrenal glands in an anencephalic foetus, the spontaneous onset of labour at or just beyond term can occur. It is of interest in this respect that in anencephalic infants, cortisol production rates measured within 24 h of birth have been reported at the lower limit of normal (Kenny, Preeyasombat, Spaulding & Migeon, 1966). Whether neonatal cortisol production rates reflect those *in utero* is not known, nor is there any information on the relationship between cortisol production rates and the length of gestation. Unfortunately Kenny *et al.* (1966) do not record the gestational age at delivery of the anencephalic infants on whom these observations were based.

Further confirmation that the adrenal gland of the anencephalic infant can synthesize corticosteroids comes from the work of Shahwan, Oakey & Stitch (1969) who demonstrated the ability of adrenal tissue from newborn anencephalic infants to convert pregnenolone and progesterone to cortisol and corticosterone *in vitro*. Since the anencephalic adrenal glands were composed almost entirely of definitive zone tissue, these authors imply that this zone must be concerned in the biosynthesis of corticosteroids in the normal newborn infant. The length of gestation in anencephaly may therefore relate more to the extent of the definitive than to the extent of the foetal zone in the adrenal glands.

Congenital adrenal hypoplasia and prolonged pregnancy

There are now several published reports of an association between primary foetal adrenal hypoplasia, in which there is no apparent pituitary abnormality, and continuation of pregnancy beyond term (O'Donohoe & Holland, 1968; Roberts & Cawdery, 1970; Fliegner, Schindler & Brown, 1972). As in anencephaly, the spontaneous onset of labour can occur, albeit well beyond the expected date of delivery, and the plasma cortisol levels in the affected infant at birth can be normal, although in the case reported by Roberts & Cawdery (1970) there was no apparent rise in plasma cortisol following injection of ACTH into the infant.

Congenital adrenal hyperplasia and the length of gestation

In infants with congenital adrenal hyperplasia one of several adrenal steroid-metabolising enzymes may be inactive and the adrenal glands, although large and hyperactive in some respects, are also functionally incompetent. The length of gestation might therefore be diminished or prolonged if foetal adrenal activity were an important determinant. A study from Cardiff (Price, Cone & Keogh, 1971) reported no difference in the length of gestation between infants with congenital adrenal hyperplasia due to a 21-hydroxylase deficiency, whether with a simple virilism or with a severe salt-losing tendency, and their normal unaffected siblings.

However, in these infants, as in anencephalic infants, cortisol production rates can be within the normal range (Kenny, Malvaux & Migeon, 1963; Visser, 1966).

Thus, from evidence in these three congenital abnormalities, namely anencephaly, adrenal hypoplasia and adrenal hyperplasia, it would not appear necessary for the human foetal adrenal to be unusually active to promote parturition, nor is it the total mass of adrenal tissue that dictates the length of gestation. Nevertheless, the data from pregnancies associated in particular with anencephaly and with foetal adrenal hypoplasia, suggest strongly that there is an association between foetal adrenal endocrine activity and the timing of the onset of labour. The mechanism by which this activity might regulate uterine activity and gestational length still remains in doubt but is being investigated in our laboratories.

The length of human pregnancy is influenced not only by the congenital abnormalities discussed above, but also by foetal adrenal hyperplasia, smoking and adverse social factors, all of which are associated with a shortening of gestational length, and placental sulphatase deficiency which can apparently delay the onset of labour beyond term.

Foetal adrenal hyperplasia and premature labour

A study of the relationship between foetal adrenal weight and the cause of premature delivery in human pregnancy was prompted by the findings of van Rensburg (1965) who reported that adrenal hyperplasia was present both in the aborted foetuses and in the does of habitually aborting Angora goats.

In a group of 79 infants delivered before the end of the 36th week of pregnancy, and who were either stillborn or died within 12 h of birth, the weights of the adrenal glands in relation to foetal weight and gestational age were determined (Anderson, Laurence, Davies, Campbell & Turnbull, 1971). This study showed that in cases of similar foetal weight and gestational age, the mean weight of the adrenals of infants delivered prematurely for unexplained reasons was approximately 1 g more than the mean weight of the adrenals of those delivered due to ante-partum haemorrhage. Histological examination of these adrenal glands showed that the increased size was neither due to haemorrhage nor to any abnormality of the gland, but rather to a true hyperplasia.

Our findings of an association between foetal adrenal hyperplasia and unexplained premature labour suggest either that in some cases there may be a causal relationship between the premature onset of labour and over-activity of the adrenal, or that both the hyperplasia and the premature

labour are the result of some other causal agent. Against the latter hypothesis is the finding that other organs, such as the kidneys and heart, are
not increased in weight.

It is interesting to surmise that in unexplained premature labour the
trigger stimulus may be an early increased activity of the hypothalamus–
pituitary–adrenal system, although there is little evidence to support this
at present, and the mode of action of drugs such as isoxsuprine and ethanol
in the prevention of premature labour would be difficult to explain on the

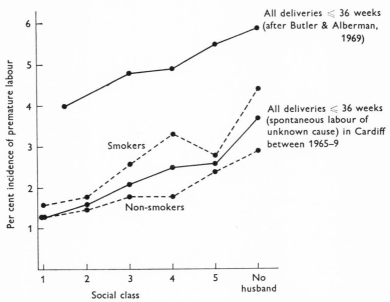

Fig. 2. The relationship between the incidence of premature onset of labour (per cent)
and social class in human pregnancy. Values for smokers and non-smokers are also
included.

basis of this hypothesis. A great deal more biochemical data on foetal
adrenal function in normal and abnormal situations in relation to the
length of pregnancy are required.

Premature labour and social class

It has been clearly shown that social class and smoking habits can influence
the length of gestation (Butler & Bonham, 1963; Butler & Alberman, 1969),
but when these correlations were examined previously, mothers were
included in whom delivery followed induction of labour by medical or
surgical means (Butler & Alberman, 1969). We thought it of interest to
re-examine the relationship of both social class and smoking habits to the
unexplained, spontaneous premature onset of labour occurring before the

36th week of pregnancy, cases of doubtful gestation length being excluded.

Figure 2 shows the incidence of spontaneous premature labour, as defined above, in relation to social class and smoking habits in all Cardiff births where the gestational age was known with reasonable certainty during the five-year period 1965–9. The percentage of these unexplained premature deliveries is shown to increase in the lower social classes whether or not the mothers smoked during pregnancy. This is similar to the trend found by Butler & Alberman (1969) for all premature deliveries occurring before the 37th week of pregnancy (Fig. 2). Thus, some factor or factors associated with poorer social class increases the propensity of women to go into labour prematurely for unexplained reasons. Whether the state of nutrition in the mother can affect uterine activity or even foetal hypo-thalamus–pituitary–adrenal activity in human pregnancy is not known. Evidence in the horse does not support this: pregnancies of 'well-fed' mares are on average 4 days shorter than those for mares on a 'maintenance ration' (Howell & Rollins, 1951). Perhaps factors other than nutritional ones affect the length of pregnancy in women of lower social status adversely and must be taken into account in any study on the cause of premature labour in human pregnancy.

Placental sulphatase deficiency and prolonged pregnancy

Recent reports suggest a correlation between the inability of the human placenta to hydrolyse sulphoconjugates and the continuation of pregnancy beyond term and/or difficulty in induction of labour (France & Liggins, 1969; Fliegner *et al.* 1972). Fliegner *et al.* (1972) point out that the clinical problems of inducing labour are similar in both placental sulphatase deficiency and in foetal adrenal hypoplasia and suggest that since low oestrogen production, as indicated by low maternal oestriol excretion, is common to the two conditions, oestrogens are more important than 'adrenal hormones' in the mechanism of the onset of labour. However, it may be that placental hydrolysis of corticosteroid conjugates is essential for the mediation of the biological effects of potent glucocorticoids and that the deficiency of the sulphatase enzyme in the placenta prevents cleavage of the sulphates of corticosteroids as well as of oestrogens. Opposing this reasoning is the evidence of Pasqualini, Cedard, Nguyen & Alsatt (1967) who found that the ability of the human term placenta to hydrolyse corticosterone 21-sulphate was limited, although not completely absent.

Placental metabolism of corticosteroids

There are many reports in the literature that both cortisol and cortisone are present in human umbilical cord plasma (Bro-Rasmussen, Buus & Trolle, 1962; Hillman & Giroud, 1965; James, 1966). The source of these steroids seems to be, at least in part, independent of the maternal adrenal glands (James, 1966), since, in the umbilical cord blood of the babies of two patients with adrenal insufficiency, both cortisol and cortisone were detected even when none was present in the maternal plasma. It seems likely that the placenta contributes to the pool of circulating cortisone in the umbilical cord blood since James (1966) found a higher concentration of cortisone in the umbilical vein than in the umbilical artery at term and, *in vivo*, Pasqualini *et al.* (1970) showed that the human placenta in mid-pregnancy is able to effect the interconversion of cortisol and cortisone. Term placental tissue has also been shown to be capable of oxidising cortisol to cortisone *in vitro* (Osinski, 1960; Sybulski & Venning, 1961), thus providing a means of disposing of biologically active cortisol and hence regulating the glucocorticoid environment of the human foeto-placental unit.

Our preliminary studies with human placentas suggest that the activity of the 11β-hydroxysteroid dehydrogenase enzyme may change during pregnancy in favour of conserving the potent glucocorticoid, cortisol, at term.

The interconversion of cortisol and cortisone in placental tissue obtained from two patients at 18 weeks of gestation who were undergoing thera-peutic abortion by hysterotomy for psychiatric reasons, has been compared with cortisol metabolism in placental tissue from two patients delivered vaginally at term after the spontaneous onset of labour with a normal foetus. Equimolar amounts (18 nmol) of [4-^{14}C]cortisone and [1,2-^{3}H]-cortisol were incubated with 1 g of minced saline-washed placental tissue from each patient for 2 h at 37 °C in Krebs–Ringer bicarbonate glucose medium. The two placental incubations at 18 weeks were pooled and so were the two at 40 weeks and, after the addition of 500 μg each of non-radioactive cortisol and cortisone to the incubation flasks, the steroids were extracted, isolated after extensive chromatography and derivative formation, and specific activities measured by procedures previously described (Anderson, 1972).

At 18 weeks of gestation, the human placenta extensively metabolised cortisol to cortisone (94·6 per cent conversion; Fig. 3) in contrast to the situation at term when only 21 per cent of the substrate cortisol was metabolised to cortisone. The metabolism of labelled cortisone at 18 weeks was minimal, whereas at 40 weeks, 40 per cent of cortisone was trans-formed to cortisol at the end of the 2 h incubation period.

These results confirm the findings of Osinski (1960) and of Sybulski &

Venning (1961) that the human placenta at term contains active 11β-hydroxysteroid dehydrogenase activity and also demonstrate an increased ability of the term placenta to metabolise cortisone to cortisol. This finding may have important implications for the onset of labour in human pregnancy if glucocorticoid activity is as vital for human parturition as it is for other species. These preliminary results need confirmation and we have yet to establish the timing of this change in placental steroid-metabolising enzyme activity, for it could be that the changes we have demonstrated occurred subsequent to the onset of labour and are not a pre-requisite for that event. (See note on p. 162.)

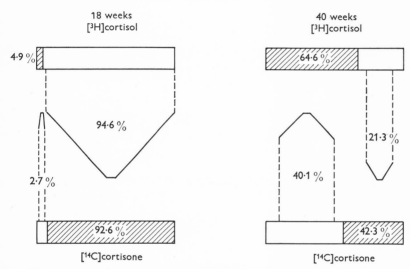

Fig. 3. The percentage interconversion of isotopically-labelled cortisol and cortisone after incubation *in vitro* for 2 h of placental tissue from two women at 18 weeks and two women at 40 weeks of gestation. The hatched sections of the horizontal bars show the percentage not metabolised.

Failure of induction of labour in human pregnancy after maternal or foetal administration of dexamethasone

In cows and sheep, labour can be initiated in late pregnancy by injection into the mother of 10–20 mg of the synthetic glucocorticoid, dexamethasone (Adams & Wagner, 1970; Fylling, 1971; Edqvist *et al.* 1972). The administration of a single dose of 25 mg dexamethasone into the foetal sheep *in utero* has also been found to be effective in the induction of parturition after 125 days of gestation (A. B. M. Anderson & A. C. Turnbull, unpublished observations). Since this represents a potentially useful means of medical induction in human pregnancy, we have attempted to initiate parturition in women at term using a single injection of dexamethasone.

Six healthy women with a normal pregnancy and a single foetus who were within a week of their expected date of delivery were given a single 20 mg intramuscular injection of dexamethasone. In no case did labour begin within 72 h of the injection, and in all women labour was induced subsequently by surgical means.

In one patient at 38 weeks of gestation with an anencephalic foetus *in utero*, 25 mg of dexamethasone were injected into the foetus using a

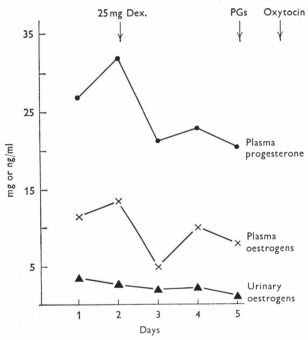

Fig. 4. Maternal plasma concentration of progesterone (ng/ml), total unconjugated oestrogens (ng/ml) and 24 h urinary oestrogens (mg/ml) in a patient with an anencephalic foetus, before and after administration of 25 mg dexamethasone (Dex.) to the foetus *in utero*. Prostaglandins (PGs) and oxytocin were administered subsequently.

transabdominal approach. Labour did not occur within 72 h of the injection and the patient was then given oral prostaglandins and intravenous oxytocin during the next few days. Induction of labour proved very difficult in this case but the patient eventually delivered an anencephalic foetus weighing 1·7 kg which, at examination *post mortem* was found to be anephric, with adrenal glands weighing only 0·3 g. No amniotic fluid was seen. The placenta was small and unhealthy-looking and weighed 200 g.

Plasma progesterone, total plasma unconjugated oestrogens (using the methods described by Symonds *et al.* 1972) and the total urinary oestrogen concentration were measured in this patient for 2 days before the intra-foetal injection of dexamethasone and for 3 days thereafter. The results are shown in Fig. 4. Although the level of all the steroids apparently fell in the days after dexamethasone administration, the trend may not be significant since the concentration of plasma oestrogens and progesterone during pregnancy shows considerable fluctuation throughout the day (Symonds *et al.* 1972). The low level of plasma oestrogens in this case of anencephaly may be related to the inability of the anencephalic adrenal glands to synthesise the oestrogen precursor, dehydroepiandrosterone sulphate (DHAS) but was probably related to the small and unhealthy-looking placenta which may have had a limited capacity for steroid synthesis, as manifested by the extremely low levels of plasma progesterone present in the mother.

From these investigations it appears that in human pregnancy, dexamethasone, whether administered to the mother or to the foetus, is not effective in the induction of labour. In this respect the findings are similar to those in the mare (Adams & Wagner, 1970; Drost, 1972) and in the guinea-pig (F. A. Harrison, personal communication). The recent findings of Honnebier & Swaab (1973) are of interest: a total dose of 21 mg of synthetic ACTH was given to an anencephalic foetus *in utero* over a period of 6 weeks. Labour did not ensue by the 42nd week of pregnancy although the foetal adrenal glands weighed over 5 g at birth and the levels of corticosteroids and of DHAS in the umbilical cord plasma were within the normal limits. It may be that the placenta of the anencephalic foetus is not as normal in terms of its ability to metabolise steroids, as has previously been assumed. Our preliminary results suggest that this could be so, since the placenta of an anencephalic foetus delivered at 45 weeks of pregnancy after induction of labour bears a greater similarity to the normal human placenta at 18 weeks than at 40 weeks in terms of its ability to interconvert cortisol and cortisone *in vitro*. Thus, although the anencephalic adrenal glands could be stimulated to produce normal amounts of cortisol as is implied from the study of Honnebier & Swaab (1973) the high activity of the 11β-hydroxysteroid dehydrogenase enzyme in the placenta may inactivate any foetal cortisol to cortisone. These results suggest that certain steroid-metabolising enzymes in the human placenta are under the control of the foetal endocrine system and are not activated in the absence of a functional foetal hypothalamus as in certain cases of anencephaly.

PROGESTERONE

In sheep

The possibility that there is a withdrawal of a 'progesterone block' on myometrial activity in late pregnancy in the sheep has not been established although it is known that levels of progesterone in uterine vein and peripheral vein fall during the last 7 to 10 days of pregnancy in this species. However, the uterine production rate of progesterone is high up to 3 days before delivery (Bedford, Challis, Harrison & Heap, 1972) and the experiments of Liggins, Grieves, Kendall & Knox (1972) led these authors to conclude that a fall in the progesterone concentration in the plasma or myometrium is not likely to be a major factor controlling the time of onset of parturition in the ewe.

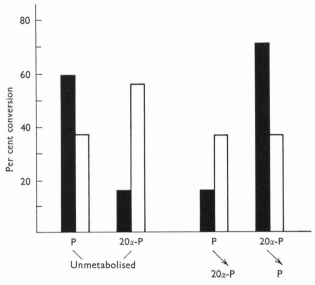

Fig. 5. Percentage interconversion of isotopically-labelled progesterone (P) and 20α-dihydroprogesterone (20α-P) after incubation of sheep myometrium for 2 h *in vitro*. Black bars = myometrium obtained at 137 days of gestation (control); white bars = myometrium obtained at 137 days of gestation following intrafoetal infusion of Synacthen for 65 h.

We have investigated the possibility that a local change in steroid-metabolising enzyme activity in the sheep myometrium might be important in the control of the proegsterone environment of that tissue; a change which may not necessarily manifest itself in dramatic changes in circulating steroid levels. Myometrial tissue from two sheep at 137 days of gestation, was incubated *in vitro* with isotopically-labelled progesterone and 20α-dihydroprogesterone and the degree of interconversion of these two steroids after a 2 h incubation period was studied. One ewe acted as a 'control'; in the other the foetus *in utero* had been infused with 250 μg Synacthen/24 h, for 65 h via an indwelling catheter in the inferior vena cava. In both sheep, Caesarean hysterectomy was performed under epidural analgesia. Equimolar amounts (20 nmol) of [4-^{14}C]progesterone and [1,2-^{3}H]20α-dihydroprogesterone were incubated with 2 g of myometrial tissue, in Krebs–Ringer bicarbonate glucose medium for 2 h at 39·5 °C. The steroids were purified, characterised, and the percentage interconversion calculated. Figure 5 shows the results. The striking feature was that the myometrium from the sheep treated with intrafoetal Synacthen metabolised more of the substrate progesterone to 20α-dihydroprogesterone but much less of the 20α-dihydroprogesterone to progesterone as compared with the 'control' myometrium. These findings suggest that, within the myometrial tissue itself, a change in the activity of the 20α-hydroxysteroid dehydrogenase enzyme could be occurring under the influence of Synacthen, a compound known to induce parturition after intrafoetal infusion for 72 h. This change in enzyme activity may represent the means by which the myometrium is released from the inhibitory effect of progesterone.

In human pregnancy

There is very little evidence that the onset of labour at term in women is associated with a fall in peripheral plasma levels of progesterone. The concentration of this steroid appears to reach its peak at parturition (Tulchinsky, Hobel, Yeager & Marshall, 1972). Nor is there evidence for an increase in the ability of the myometrium to inactivate progesterone; the results of Zander et al. (1969) suggest that the activity of the 20α-hydroxysteroid dehydrogenase enzyme in human myometrial tissue increasingly favours the oxidative reaction as term approaches.

We have investigated the possibility that human placental tissue might develop an enhanced capacity to convert progesterone to 20α-dihydroprogesterone during gestation but to date have found no evidence to support this theory.

If the situation in ovine pregnancy whereby increasing foetal adrenal cortisol production triggers a chain of endocrine events which lead to parturition is analogous to that in human pregnancy, and if a withdrawal of progesterone is in some way involved in this chain of events, then more sophisticated experimental designs will be required for investigations of human pregnancy than have been previously employed.

OESTROGENS

In sheep

The demonstration by Challis (1971) that there is a very rapid increase in circulating maternal oestrogens in sheep within 2 days of lambing has stimulated several studies on the role of oestrogens in parturition in the ewe (Bedford *et al.* 1972; Liggins *et al.* 1972; Thorburn *et al.* 1972). It is known that a synthetic oestrogen like stilboestrol will induce labour if administered to the ewe near term. At term, oestrone is quantitatively the major circulating oestrogen in sheep; oestradiol-17β concentrations were found to be about half those of oestrone (Thorburn *et al.* 1972).

In human pregnancy

The relative proportion of unconjugated oestrogens in the peripheral plasma of pregnant women in late pregnancy is the reverse of that found in sheep at term. In human pregnancy at term, oestradiol-17β and oestriol are quantitatively the most important circulating oestrogens, with oestrone present in approximately half their concentration (Tulchinsky *et al.* 1972). The pattern of the increase during gestation in these oestrogens in human pregnancy is interesting in that, after approximately the 32nd week of pregnancy, the concentration of oestradiol-17β and of oestriol rises sharply, whereas the concentration of plasma oestrone flattens off in the last 8 weeks.

A few years ago we measured the urinary excretion of oestriol and oestrone at the 34th week of pregnancy, and found that in patients with a high level of uterine activity and an early onset of labour, oestriol excretion was high and oestrone low (Turnbull, Anderson & Wilson, 1967). In contrast, the lower the urinary output of oestriol or the higher the oestrone at this stage of pregnancy, the more prolonged the pregnancy. At that time we suggested that, since the foetus is known to play a vital part in the intermediary metabolism of oestriol, it seemed likely that the endocrine activity in the foetus, and in particular adrenal activity, in some way regulated the level of uterine contractility and the duration of pregnancy.

The results of the investigations on the urinary excretion of oestrone (Turnbull *et al.* 1967) and the pattern of increase in maternal plasma levels of oestrone and oestradiol-17β (Tulchinsky *et al.* 1972) suggest that the placenta may also be playing a regulatory role in timing the onset of labour, since these findings might indicate increasing conversion by the placenta of biologically inactive oestrone to the biologically active oestradiol-17β with advancing gestation. We are at present investigating this possibility, although the relative proportions of these oestrogens in maternal plasma could merely be reflecting differences in the interconversion of oestrone and oestradiol-17β in the mother's peripheral circulation.

Whether oestrogens in physiological amounts can influence uterine activity in human pregnancy is uncertain although pharmacological doses of oestradiol benzoate given to women in late pregnancy increase the contractility of the uterus (Järvinen, Luukkainen & Väistö, 1965) and maternally-administered oestradiol-17β hemisuccinate increases the sensitivity of the uterus to oxytocin (Pinto, Lerner, Glauberman & Pontelli, 1966). The exact role of oestrogens in relation to parturition in human pregnancy remains to be determined.

PROSTAGLANDINS

In sheep

There is evidence that prostaglandin $F_{2\alpha}$ ($PGF_{2\alpha}$) may be involved in the control of ovine parturition (Liggins & Grieves, 1971) although the infusion of $PGF_{2\alpha}$ at a rate of up to 16 μg/min into the aorta of a ewe at 134 days of gestation did not induce uterine activity (Liggins *et al.* 1972).

Since the foetus seems to be critically involved in the onset of labour in sheep (Liggins, 1968) we have investigated the possibility that intra-foetal rather than maternal prostaglandins might have a more potent stimulatory effect on the myometrium (Keirse, Patten, Anderson, Turnbull, Johns, Wooster & Pickles, 1973). Accordingly, in three sheep under epidural analgesia at 118, 120 and approximately 140 days of pregnancy, catheters were inserted into the inferior vena cava of the foetus *in utero*, the catheters were brought out through the flank of the ewe and connected to a constant infusion apparatus (Palmer Ltd, London). In the ewe at 140 days, a catheter was also inserted into a uterine artery and again, brought out through the flank. Amniotic fluid pressure was recorded in each sheep by means of an open-ended polyethylene catheter inserted into the amniotic sac, brought out through the skin incision and connected to a pressure transducer and pen recorder (Sanborn Company, Mass., USA).

As shown in Table 1, at 118 and 120 days of gestation, PGE_1 and $PGF_{2\alpha}$ were infused into the foetal inferior vena cava at rates of up to 22·5 μg $PGF_{2\alpha}$/min for 59 h and 16·6 μg PGE_1/min for 30 h. At term, 5 μg PGE_1/min was infused first into the foetus for 5 h, then into the maternal uterine artery for a further 5 h, with an interval of 2 h between each infusion. In these three experiments, labour as indicated by amniotic fluid pressure recordings or delivery of the foetus did not ensue.

Although prostaglandins of the E series have not been found in the pregnant sheep (Liggins & Grieves, 1971) the ability of the non-pregnant ovine uterus to synthesise them has been demonstrated (Nugteren, Beerthuis & Van Dorp, 1966). It seems unlikely that the lack of response

Table 1. *Infusion of prostaglandins at different rates and for different periods into three sheep*

(From Keirse, Patten, Anderson, Turnbull, Johns, Wooster & Pickles, 1973)

Gestation (days)	Prostaglandin infusion	Dose (μg/min)	Time infused (h)	Total dose (mg)
118	F$_{2\alpha}$ (intrafoetal)	5·6*	3	
		5·6	8	
		11·2	12	60
		22·5	36	
120	E$_1$ (intrafoetal)	2·5*	2	
		2·5	4	13
		5·0	8	
		8·3	12	
140 (approx.)	E$_1$ (intrafoetal)	5·0*	5	
		Infusion stopped for 2 h		3
	E$_1$ (maternal uterine artery)	5·0*	5	

* The upper limit of a gradually increasing rate of infusion.

of the sheep myometrium to PGF$_{2\alpha}$ was due to insufficient dosage since the secretion rate of PGF$_{2\alpha}$ into the uterine vein at term has been estimated to be of the order of 16 μg/min (Challis *et al.* 1972). Other reasons for the lack of response to the infused prostaglandins might include their rapid metabolism or the failure of the ovine myometrium to respond to prostaglandins until other endocrine events leading to the onset of labour have occurred.

To determine whether the pregnant sheep myometrium was inherently capable of responding to prostaglandins, experiments *in vitro* were carried out. The observations were made on strips of myometrium taken from six anaesthetised ewes between 120 days of pregnancy and term (145 days). The strips were mounted in 5 or 10 ml organ baths maintained at 37 °C and bathed with a Krebs–Heneleit type solution through which was bubbled a mixture of 5 per cent CO$_2$ and 95 per cent O$_2$. The contractions of the strips were recorded isotonically on smoked drums, or isometrically by means of strain gauges and pen recorders. Prostaglandins were dissolved in a dilute solution of NaHCO$_3$ to give a neutral pH, and made up to a stock solution of 100 μg/ml. Volumes of not more than 0·1 ml were added to the organ baths by means of microsyringes.

Preparations from all but one of the ewes were responsive to prostaglandins in as low a concentration as 10 ng PGE/ml or 50 ng PGF$_{2\alpha}$/ml. These observations indicate that pregnant sheep myometrium responds to prostaglandins in a manner similar to pregnant human myometrium (Embrey & Morrison, 1968). The tissue is usually adequately sensitive to the stimulatory action of PGE$_1$ and PGF$_{2\alpha}$, and failure to induce labour by infusion of these compounds *in vivo* must be due to some factor other than myometrial insensitivity to prostaglandins.

In contrast to the failure of prostaglandins to effect contractions of sheep myometrium *in vivo*, the infusion of oxytocin (1 to 32 mu./min) either into the foetal inferior vena cava or into the maternal jugular vein

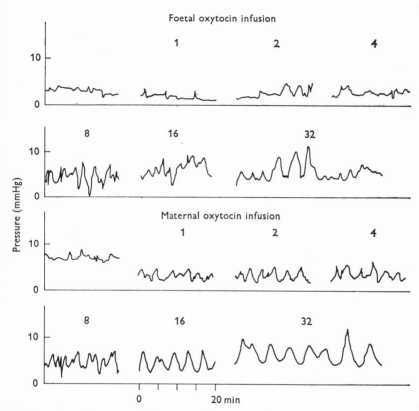

Fig. 6. Segments of intra-amniotic pressure (mmHg) recordings in a pregnant ewe infused with oxytocin at 1, 2, 4, 8, 16 and 32 mu./min, first into the foetal inferior vena cava, then into the maternal jugular vein.

at 125 days of gestation stimulated the myometrium to contract (Fig. 6). The pattern of uterine response seemed to depend on the route of administration, in that the amniotic fluid pressure changes appeared more regular and co-ordinated when the oxytocin was given via the maternal jugular vein.

The results of experiments with prostaglandins and oxytocin in sheep represent the antithesis to findings in human pregnancy where, in midgestation, prostaglandins are very much more effective than oxytocin in causing effective uterine activity.

In human pregnancy

In a recent review article Karim (1972) has discussed the evidence in support of the suggestion that prostaglandins have a role in spontaneous abortion and labour in human pregnancy.

Prostaglandins were first observed in human amniotic fluid by Karim & Devlin (1967). Prostaglandins of the E series, mainly PGE_1, were found in late pregnancy and in labour while $PGF_{2\alpha}$ appeared during labour. These early measurements, however, were obtained with a biological assay, a method which is sensitive but lacks specificity (Horton, 1972). Using a modification of a method specific for the determination of prosta-

Table 2. *Prostaglandin E_2 (PGE_2) in human amniotic fluid at term*

(From Keirse & Turnbull, unpublished results)

Patient	Labour	Cervical dilatation (cm)	PGE_2 (ng/ml)
1	Not in labour	—	<0·2
2	Not in labour	—	<0·5
3	Not in labour	—	<0·5
4	Not in labour	—	<0·5
5	Induced labour	4	4·0
6	Spontaneous	4	5·0
7	Spontaneous	5	4·7
8	Spontaneous	5	5·1
9	Induced labour	7	12·1
10	Spontaneous	Full	11·0

glandins of the E series (Jouvenaz, Nugteren, Beerthuis & Van Dorp, 1970), the PGE content of human amniotic fluid has been measured in late pregnancy and labour (Keirse and Turnbull 1973, in preparation). Preliminary findings showed that neither PGE_1 nor PGE_2 were detectable in any amniotic fluid sample obtained before labour (Table 2). All samples during labour, however, contained PGE_2 in amounts varying from 4·0 to 12·1 ng/ml, the highest concentrations being found later in labour. Table 2 shows the cervical dilatation at the time of sampling of the amniotic fluid and the trend for the PGE_2 levels to increase as dilatation advances. There appeared to be no gross differences between PGE_2 concentrations in oxytocin-induced and spontaneous labour. PGE_1 was not detected in any sample during labour.

These results suggest that PGE_2 is present in amniotic fluid as a consequence of labour rather than as a factor in its initiation. However, further studies are clearly needed and the temporal relationships between the synthesis of prostaglandins in tissues and their appearance in the amniotic fluid have yet to be established.

CONCLUSION

The development of highly sensitive and specific radioimmunoassays for the measurement of plasma and tissue concentrations of steroid and protein hormones and of prostaglandins should enable a more rapid advance of knowledge in the field of human parturition over the next few years. Although the factors determining the onset of labour in human pregnancy may prove difficult to enumerate with absolute certainty, a great deal of useful knowledge can be gained from carefully selected and collected clinical material. It must be hoped that the experimental evidence obtained from studies on other species will enable us to control uterine activity in human pregnancy more physiologically than is possible at present.

ACKNOWLEDGMENTS

We wish to thank Dr R. G. Jacomb of Upjohn Limited, Crawley, Sussex and Professor D. A. Van Dorp of Unilever Laboratories, Vlaardingen, Holland, for their generous gifts of PGE_1 and $PGF_{2\alpha}$. We are indebted to Mr J. Wilson and the staff of the Dr Len West Animal Research Laboratory, Sully Hospital, for expert assistance with the animals. Part of the work reported in this paper has been financed by the Wellcome Trust and the Medical Research Council. The Tenovus Organisation in Cardiff generously provided laboratory facilities for some of the studies on sheep.

REFERENCES

ADAMS, W. M. & WAGNER, W. C. (1970). The role of corticoids in parturition. *Biol. Reprod.* **3**, 223–228.

ANDERSON, A. B. M. (1972). Some aspects of steroid metabolism in adult and foetal sheep. Ph.D. thesis, University of Wales.

ANDERSON, A. B. M., LAURENCE, K. M., DAVIES, K., CAMPBELL, H. & TURNBULL, A. C. (1971). Fetal adrenal weight and the cause of premature delivery in human pregnancy. *J. Obstet. Gynaec. Br. Commonw.* **78**, 481–488.

ANDERSON, A. B. M., LAURENCE, K. M. & TURNBULL, A. C. (1969). The relationship in anencephaly between the size of the adrenal cortex and the length of gestation. *J. Obstet. Gynaec. Br. Commonw.* **76**, 196–199.

ANDERSON, A. B. M., PIERREPOINT, C. G., GRIFFITHS, K. & TURNBULL, A. C. (1972). Steroid metabolism in the adrenals of fetal sheep in relation to natural and corticotrophin-induced parturition. *J. Reprod. Fert.* Suppl. **16**, 25–37.

BEDFORD, C. A., CHALLIS, J. R. G., HARRISON, F. A. & HEAP, R. B. (1972). The role of oestrogens and progesterone in the onset of parturition in various species. *J. Reprod. Fert.* Suppl. **16**, 1–23.

BRO-RASMUSSEN, F., BUUS, O. & TROLLE, D. (1962). Ratio cortisone/cortisol in mother and infant at birth. *Acta endocr., Copenh.* **40**, 579–583.

BUTLER, N. R. & ALBERMAN, E. C. (eds.) (1969). *Perinatal problems*, pp. 49–51. Edinburgh & London: E. & S. Livingstone Ltd.

BUTLER, N. R. & BONHAM, D. G. (eds.) (1963). *Perinatal mortality*, pp. 126–128. Edinburgh & London: E. & S. Livingstone Ltd.

CHALLIS, J. R. G. (1971). Sharp increase in free circulating oestrogens immediately before parturition in sheep. *Nature, Lond.* **229**, 208.

CHALLIS, J. R. G., HARRISON, F. A., HEAP, R. B., HORTON, E. W. & POYSER, N. L. (1972). A possible role of oestrogens in the stimulation of prostaglandin $F_{2\alpha}$ output at the time of parturition in a sheep. *J. Reprod. Fert.* **30**, 485–488.

COMERFORD, J. B. (1965). Pregnancy with anencephaly. *Lancet*, i, 679–680.

DROST, M. (1972). Failure to induce parturition in pony mares with dexamethasone. *J. Am. vet. med. Ass.* **160**, 321–322.

EDQVIST, L.-E., EKMAN, L., GUSTAFSSON, B., JACOBSSON, S.-O., JOHANSSON, E. D. B. & LINDELL, J.-O. (1972). Peripheral plasma levels of oestrone and progesterone in pregnant cows treated with dexamethasone. *Acta endocr., Copenh.* **71**, 731–742.

EMBREY, M. P. & MORRISON, D. C. (1968). The effect of prostaglandins on human pregnant myometrium *in vitro*. *J. Obstet. Gynaec. Br. Commonw.* **75**, 829–832.

FLIEGNER, J. R. H., SCHINDLER, I. & BROWN, J. B. (1972). Low urinary oestriol excretion during pregnancy associated with placental sulphatase deficiency or congenital adrenal hypoplasia. *J. Obstet. Gynaec. Br. Commonw.* **79**, 810–815.

FRANCE, J. T. & LIGGINS, G. C. (1969). Placental sulfatase deficiency. *J. clin. Endocr.* **29**, 138–141.

FYLLING, P. (1971). Premature parturition following dexamethasone administration to pregnant ewes. *Acta endocr., Copenh.* **66**, 289–295.

HILLMAN, D. A. & GIROUD, C. J. P. (1965). Plasma cortisone and cortisol levels at birth and during the neonatal period. *J. clin. Endocr. Metab.* **25**, 243–248.

HONNEBIER, W. J. & SWAAB, D. R. (1973). The role of the human foetal brain in the onset of labour. *J. Endocr.* **57**, xxx–xxxi.

HORTON, E. W. (1972). *Prostaglandins*, Monographs on Endocrinology vol. 7. London: William Heinemann Medical Books.

HOWELL, C. E. & ROLLINS, W. C. (1951). Environmental sources of variation in the gestation length of the horse. *J. Anim. Sci.* **10**, 789–796.

JAMES, V. H. T. (1966). Corticosteroid secretion by human placenta and foetus. *Eur. J. Steroids*, **1**, 5–14.

JÄRVINEN, P. A., LUUKKAINEN, T. & VÄISTÖ, L. (1965). The effect of oestrogen treatment on myometrial activity in late pregnancy. *Acta obstet. gynec. scand.* **44**, 258–264.

JOUVENAZ, G. H., NUGTEREN, D. H., BEERTHUIS, R. K. & VAN DORP, D. A. (1970). A sensitive method for the determination of prostaglandins by gas chromatography with electron-capture detection. *Biochem. biophys. Acta*, **202**, 231–234.

KARIM, S. M. M. (1972). Physiological role of prostaglandins in the control of parturition and menstruation. *J. Reprod. Fert.* Suppl. **16**, 105–119.

KARIM, S. M. M. & DEVLIN, J. (1967). Prostaglandin content of amniotic fluid during pregnancy and labour. *J. Obstet. Gynaec. Br. Commonw.* **74**, 230–234.

KEIRSE, M. J. N. C., PATTEN, P. T., ANDERSON, A. B. M., TURNBULL, A. C., JOHNS, A., WOOSTER, M. J. & PICKLES, V. R. (1973). Pregnant sheep myometrium responds to prostaglandins *in vitro* but not *in vivo*. *Int. Res. Comm. Syst.*, April.

KENNY, F. M., MALVAUX, P. & MIGEON, C. J. (1963). Cortisol production rate in newborn babies, older infants and children. *Pediatrics*, **31**, 360–373.

KENNY, F. M., PREEYASOMBAT, C., SPAULDING, J. S. & MIGEON, C. J. (1966). Cortisol production rate. IV. Infants born of steroid-treated mothers and of diabetic mothers. Infants with trisomy syndrome and with anencephaly. *Pediatrics*, **37**, 960–966.

LIGGINS, G. C. (1968). Premature parturition after infusion of corticotrophin or cortisol into foetal lambs. *J. Endocr.* **42**, 323–329.

LIGGINS, G. C. (1969). The foetal role in the initiation of parturition in the ewe. In *Foetal Autonomy. Ciba Fdn. Symp.* eds. G. E. W. Wolstenholme & M. O'Connor, pp. 218–237. London: Churchill.

LIGGINS, G. C. & GRIEVES, S. A. (1971). Possible role for prostaglandin $F_{2\alpha}$ in parturition in sheep. *Nature, Lond.* **232**, 629–631.

LIGGINS, G. C., GRIEVES, S. A., KENDALL, J. Z. & KNOX, B. S. (1972). The physiological roles of progesterone, oestradiol-17β and prostaglandin $F_{2\alpha}$ in the control of ovine parturition. *J. Reprod. Fert.* Suppl. **16**, 85–103.

LIGGINS, G. C., HOLM, L. W., & KENNEDY, P. C. (1966). Prolonged pregnancy following surgical lesions of the foetal lamb pituitary. *J. Reprod. Fert.* **12**, 149.

MALPAS, P. (1933). Postmaturity and malformations of the foetus. *J. Obstet. Gynaec. Br. Emp.* **40**, 1046–1053.

MILIC, A. B. & ADAMSONS, K. (1969). The relationship between anencephaly and and prolonged pregnancy. *J. Obstet. Gynaec. Br. Commonw.* **76**, 102–112.

NUGTEREN, D. H., BEERTHUIS, R. K. & VAN DORP, D. A. (1966). The enzymic conversion of *all-cis* 8,11,14-eicosatrienoic acid into prostaglandin E_1. *Recl. Trav. chim. Pays-Bas Belg.* **85**, 405–419.

O'DONOHOE, N. V. & HOLLAND, P. D. J. (1968). Familial congenital adrenal hypoplasia. *Archs Dis. Childh.* **43**, 717–723.

OSINSKI, P. A. (1960). Steroid 11β-ol dehydrogenase in human placenta. *Nature, Lond.* **187**, 777.

PASQUALINI, J. R., CEDARD, L., NGUYEN, B. L. & ALSATT, E. (1967). Differences in the activity of human term placenta sulphatases for steroid ester sulphates. *Biochim. Biophys. Acta*, **139**, 177–179.

PASQUALINI, J. R., NGUYEN, B. L., UHRICH, F., WIQVIST, N. & DICZFALUSY, E. (1970). Cortisol and cortisone metabolism in the human foeto-placental unit at midgestation. *J. Steroid Biochem.* **1**, 209–219.

PINTO, R. M., LERNER, U., GLAUBERMAN, M. & PONTELLI, H. (1966). Influence of estradiol-17β upon the oxytocic action of oxytocin in the pregnant human uterus. *Am. J. Obstet. Gynec.* **96**, 857–862.

PRICE, H. V., CONE, B. A. & KEOGH, M. (1971). Length of gestation in congenital adrenal hyperplasia. *J. Obstet. Gynaec. Br. Commonw.* **78**, 430–434.

REA, C. (1898). Prolonged gestation, acrania monstrosity and apparent placenta previa in one obstetrical case. *J. Am. med. Ass.* **30**, 1166–1167.

ROBERTS, G. & CAWDERY, J. E. (1970). Congenital adrenal hypoplasia. *J. Obstet. Gynaec. Br. Commonw.* **77**, 654–656.

SHAHWAN, M. M., OAKEY, R. E. & STITCH, S. R. (1969). Corticosteroid synthesis *in vitro* by adrenal tissue from newborn anencephalic infants. *J. Endocr.* **44**, 557–566.

SYBULSKI, S. & VENNING, E. H. (1961). The possibility of corticosteroid production by human and rat placental tissue under *in vitro* conditions. *Can. J. Biochem.* **39**, 203–214.

SYMONDS, E. M., FAHMY, D., MORGAN, C., ROBERTS, G., GOMERSALL, C. R. & TURNBULL, A. C. (1972). Maternal plasma oestrogen and progesterone levels during therapeutic abortion induced by intra-amniotic injection of prostaglandins F_{2z}. *J. Obstet. Gynaec. Br. Commonw.* **79**, 976–980.

THORBURN, G. D., NICOL, D. H., BASSETT, J. M., SHUTT, D. A. & COX, R. I. (1972). Parturition in the goat and sheep: changes in corticosteroids, progesterone, oestrogens and prostaglandin F. *J. Reprod. Fert.* Suppl. **16**, 61–84.

TULCHINSKY, D., HOBEL, C. J., YEAGER, E. & MARSHALL, J. R. (1972). Plasma estrone, estradiol, estriol, progesterone and 17-hydroxyprogesterone in human pregnancy. *Am. J. Obstet. Gynec.* **112**, 1095–1100.

TURNBULL, A. C., ANDERSON, A. B. M. & WILSON, G. R. (1967). Maternal urinary oestrogen excretion as evidence of a foetal role in determining gestation at labour. *Lancet*, ii, 627–629.

VAN RENSBURG, S. J. (1965). Adrenal function and fertility. *Jl S. Afr. vet. med. Ass.* **36**, 491–500.

VISSER, H. K. A. (1966). The adrenal cortex in childhood. Part 2: Pathological aspects. *Archs. Dis. Childh.* **41**, 113–136.

ZANDER, J., HOLZMANN, K., VON MÜNSTERMANN, A. M., RUNNEBAUM, B. & SIEBER, W. (1969). New results on the metabolism of progesterone in the foeto–placental unit. In *The Foeto–Placental Unit*, eds. A. Pecile & C. Finzi, pp. 162–175. Amsterdam: Excerpta Medica Foundation.

NOTE ADDED IN PROOF

Studies done recently in our laboratory have shown that the difference found in the metabolism of cortisol and cortisone by the human placenta at 18 and 40 weeks of gestation (Fig. 3) may have been due in part to a difference in the amount of 'foetal' and 'maternal' placenta incubated. Although we find that the foetal surface of the human placenta has a decreased ability to metabolise cortisol to cortisone, following the onset of labour at term, it cannot convert cortisone to cortisol whereas the maternal surface of the placenta and the decidua are both active in the metabolism of cortisone to cortisol.

PARTURITION IN RABBITS AND RATS

By ANNA-RIITTA FUCHS

In front of a predominantly clinical audience, I always find myself feeling somewhat apologetic when talking about rats and rabbits. To students of the physiology and endocrinology of human parturition, it might seem futile to concern oneself with the elucidation of factors that determine the length of gestation and the onset of labour in such species. In these multiparous animals with a bicornuate uterus many physiological mechanisms involved in parturition are of necessity different from those in the primates. But, however different the anatomical features of different species may be, certain mechanisms of reproductive physiology may be of common biological significance to all mammals.

Rats and rabbits offer many advantages to the investigator, besides being relatively cheap. Most important is the possibility of manipulating both physiological and endocrine factors at will, in order to assess their relative significance in parturition, a task that is almost impossible in man. Although results obtained in one species cannot be assumed to apply to another without great caution, they may at least serve as useful reminders of the multitude of ways in which nature can solve similar problems.

After this 'speech for the defence', I should like to describe first the normal course of events during and preceding parturition in these two species. One aspect of gestation and parturition is certainly common to all mammals, namely, the fact that uterine function has to be regulated in two ways: first, to accommodate the growing foetus, and secondly, to expel it in due time. I shall, therefore, devote most of my time to various aspects of uterine function in gestation and during parturition. Thereafter, I should like to examine to what extent uterine contractions alone determine the termination of gestation, and to what extent other factors, such as the functional capacity of the corpus luteum as well as the maturation of the foetuses, are involved in the process of parturition.

In the rabbit, uterine activity is slight to moderate at oestrus. After mating there is a very considerable increase in uterine activity on the first two days (Fuchs, 1972 b). A gradual decline takes place on days 3 to 5, and complete quiescence is observed from about day 7 or 8 to day 28 or 29. The uterine response to oxytocin decreases as early as days 2 and 3 and is virtually absent from days 4 to 25–28. Close to term, in the last 48 h of gestation, spontaneous activity increases slowly, and single contractions appear at long intervals. Gestation is terminated by the abrupt expulsion of the litter within 15 to 30 min. This is achieved by a series of strong, frequent uterine contractions (Fig. 1) which begin very suddenly but disappear gradually over 60 to 90 min. After parturition the uterus remains quiescent for about 12–18 h before rhythmic activity reappears (Fuchs, 1964 a).

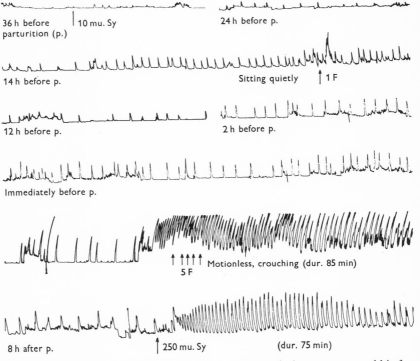

Fig. 1. Intrauterine pressure changes recorded continuously in a pregnant rabbit from 36 h before parturition (p.) until 10 h after delivery. On day 28 of gestation, one foetus from one uterine horn was removed and a recording balloon substituted in its place. The arrows denote either the delivery of one living young (F) or an injection of Syntocinon (Sy).

In the rabbit the control of myometrial function is fairly well understood and depends largely on the level of circulating ovarian hormones (Knaus, 1930 a, b, c; Makepeace, Corner & Allen, 1936; Reynolds, 1949; Csapo, 1956 a, b). Progesterone seems to be the agent responsible for the maintenance of uterine quiescence during gestation in the rabbit (Bengtsson, 1957; Schofield, 1957; Takeda, 1965). Low levels of oestrogens are also essential for the maintenance of gestation, as shown by the interruption of pregnancy after destruction of ovarian follicles by cautery (Westman,

1934) or by X-ray irradiation (Keyes & Nalbandov, 1967). In the rabbit, the principal target for ovarian oestrogens during gestation seems to be the corpora lutea (Robson, 1937), in that oestrogen is essential for the maintenance of their function. Production of progesterone by the irradiated ovaries ceases and abortion ensues unless oestrogen replacement therapy is given (Keyes & Armstrong, 1968). The action of oestrogen on the uterus is less important during gestation. In ovariectomised animals, progesterone alone can maintain gestation although foetal survival is improved by the addition of small amounts of oestrogen to the replacement regimen (Herod, Morier, Lawless & Lanman, 1972).

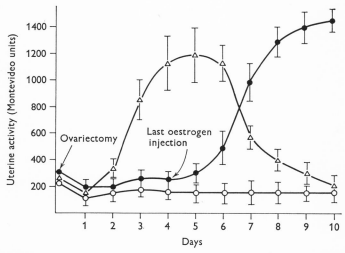

Fig. 2. Evolution of spontaneous uterine activity in rabbits after ovariectomy and the effect of oestrogen treatment on this activity. Intrauterine pressure changes were recorded by means of rubber balloons. The activity was measured in Montevideo units. △ = no treatment; ○ = 20 μg oestradiol benzoate in oil/day, i.m.; ● = oestrogen treatment and uterine distension. The activity became intense after the oestrogen treatment was discontinued. (Printed with permission from Coutinho & de Mattos, 1968.)

Normal parturition, and by that I mean the orderly evacuation of the uterus resulting in living offspring, and not merely a random emptying of the uterine contents, is however, dependent on the influence of oestrogens. In rabbits ovariectomised during gestation and maintained with progesterone, the onset of parturition after cessation of progesterone therapy is delayed and parturition itself is prolonged and often associated with retained placentas or foetuses (Csapo & Lloyd-Jacob, 1962). After ovariectomy without any replacement therapy, when both steroid hormones are withdrawn, uterine activity also develops differently from that seen at normal term. Spontaneous activity appears within 12–18 h of removal of the ovaries and is much more frequent and intense than that seen at normal term (A.-R. Fuchs, unpublished observations). Usually, the foetuses are expelled one by one at long intervals, which is in contrast to the rapid expulsion of a whole litter during normal parturition, and the incidence of stillbirth is high in ovariectomised pregnant rabbits. Oestrogen treatment suppresses the strong contractile activity present in ovariectomised rabbits as shown by Coutinho & de Mattos (1968) (Fig. 2). While suppression of uterine activity in rabbits by progesterone is accompanied by simultaneous suppression of the uterine responsiveness to oxytocin stimulation, oestrogen treatment enhances oxytocin sensitivity and increases the uterine response to oxytocin.

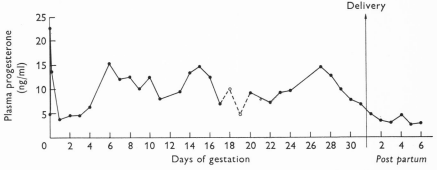

Fig. 3. Progesterone concentrations in serial samples of peripheral plasma from pregnant rabbits determined by the competitive protein-binding method. ● = mean of 4 to 6 rabbits; ○ = single determinations. (From C. Beling, A.-R. Fuchs & C. Florencio, 1973, unpublished observations.)

Determinations of ovarian progestin production during gestation by Hilliard, Spies & Sawyer (1968) and of peripheral plasma concentrations by ourselves (Beling, Fuchs & Florencio, 1973) support the concept of 'the progesterone block' first suggested by Knaus and later developed by Csapo in several publications. As can be seen in Fig. 3, the plasma progesterone levels decline before parturition and reach almost non-pregnant levels on the day before parturition.

However, even in intact rabbits at term, the withdrawal of progesterone alone is not sufficient to ensure normal parturition without neurohypophysial activation. This was demonstrated by inhibiting the release of neurohypophysial hormones by the continuous administration of ethanol (Fuchs, 1966). Parturition was postponed by 36 h on the average, and when expulsion of the litter finally took place, it followed an abnormal pattern. Contractions of the kind associated with oxytocin release were absent and the duration of delivery was much prolonged. On the other hand, injections of oxytocin given to ethanol-treated rabbits resulted in normal delivery.

Knaus was already of the opinion that withdrawal of progesterone was not enough in itself to cause labour contractions in rabbits. He suggested that the hypertrophy of myometrial muscle cells caused by distension of the gravid uterus results in improved conduction of excitatory impulses, and that this is a necessary condition for efficient labour contractions in rabbits. Our findings support Knaus's theory. Both in pregnant (Fuchs & Fuchs, 1960) and in ovariectomised animals (Fuchs & Vieira Lopes, 1970)

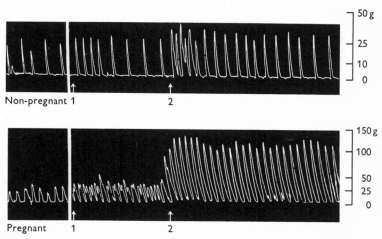

Fig. 4. Simultaneous isometric recording of uterine activity of both horns of a unilaterally pregnant rabbit 22 h *post partum*. 1 = 5 mu. oxytocin, i.v.; 2 = 50 mu. oxytocin, i.v. (Printed with permission from Fuchs & Fuchs, 1960.)

the distension of one horn greatly enhanced the response to injected oxytocin. Figure 4 illustrates this effect in unilaterally pregnant animals in which one horn has been greatly distended by the growing foetuses.

The gradual distension of the uterus by the growing conceptuses alone does not, however, suffice to elicit labour contractions without additional stimulation, which is systemic in nature as shown by simultaneous recordings of intrauterine pressure in both uterine horns of unilaterally pregnant rabbits during parturition. Before labour there was often somewhat more spontaneous activity in the pregnant horn than in the non-pregnant horn, but the expulsion of the litter was achieved with a sudden burst of uterine contractions which appeared simultaneously in both horns (compare Fig. 5 with Fig. 1). Experiments done by Gorm Wagner (1969) with the elegant technique of recording uterine contractions from a piece of myometrium transplanted into the ear, also confirm that agents released into the systemic circulation are responsible for the contractions of the uterus during parturition in the rabbit. Using a bioassay for oxytocin-like activity, investigation of blood taken during labour showed relatively high values in the samples collected at the beginning of the recorded

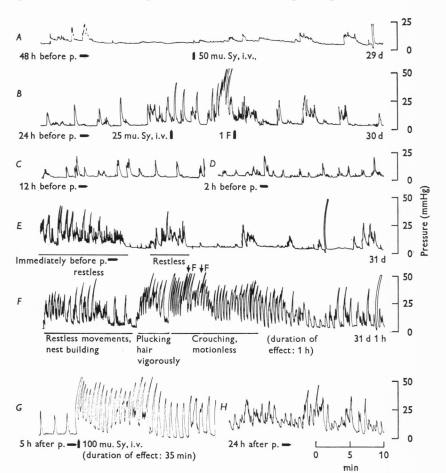

Fig. 5. Intrauterine pressure changes in the nonpregnant horn of a unilaterally pregnant rabbit recorded from 48 h before parturition (p.) until 24 h after delivery. Each arrow denotes the delivery of one living young (F). Oxytocin (Sy, Syntocinon) was injected at intervals.

series of uterine contractions and low or no activity in the samples collected later, during the diminishing phase (Fuchs, 1964 b).

Prostaglandins of uterine origin might be responsible for the initiation of labour contractions. However, the observation that indomethacin, a powerful inhibitor of prostaglandin synthesis in all tissues, had no effect on parturition when administered to pregnant and parturient rabbits (L. Speroff, 1972, personal communication) speaks strongly against this assumption.

The rat is like the rabbit in that ovarian steroids are indispensable for the maintenance of gestation. Ovariectomy during gestation leads to inhibition of implantation, resorption, retention of foetuses and placentas, abortion or prolonged and delayed parturition, according to the stage of gestation when it is performed. The corpora lutea of pregnancy are responsible for progesterone production during gestation in the rat as in the rabbit. Figure 6 shows peripheral plasma progesterone concentrations

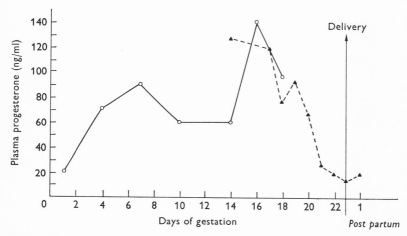

Fig. 6. Progesterone concentrations in the peripheral plasma of pregnant rats. ○ = values from Fuchs, Mok & Sundaram (1972); ▲ = values from Csapo & Wiest (1969).

throughout gestation in the rat. The pattern is similar to that observed in pregnant rabbits; the progesterone levels fall significantly after the 19th day and reach low values on days 21 and 22 (Csapo & Wiest, 1969; Fuchs, Mok & Sundaram, 1972). Schofield (1957) has suggested that in species like the rat and the rabbit, where ovarian steroids are essential throughout gestation, the mechanism controlling uterine function is similar, whereas in species where the placenta is responsible for the endocrine functions necessary for the maintenance of gestation, the mechanism might be different, and more dependent on uterine distension and a direct placental effect on the surrounding myometrium. However, recordings of uterine activity throughout gestation and delivery in the rat and the rabbit showed some important differences between these species (Fuchs, 1969b). While in the rabbit, the uterus is completely quiescent until about 24 to 28 h before parturition and does not respond even to large doses of oxytocin, the uterus of the pregnant rat shows frequent, albeit irregular, intrauterine pressure changes throughout most of gestation and responds to moderate doses of oxytocin (10 mu.) at all stages of gestation. About 24 to 36 h before parturition, the uterus becomes more quiescent. In this period prior to parturition, the infrequent contractions gradually become stronger and more co-ordinated. Shortly before parturition a marked increase in the uterine response to oxytocin also occurs in the rat (Fuchs, 1969 b; Fuchs &

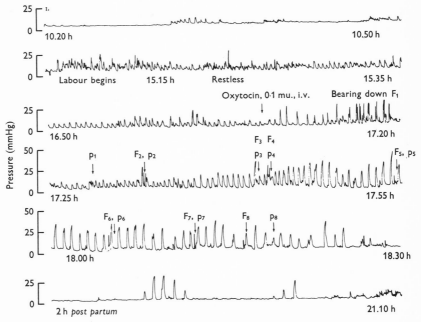

Fig. 7. Intrauterine pressure changes recorded in a pregnant rat before and during spontaneous parturition on day 22. One foetus was removed and a recording balloon substituted in its place. F = expulsion of live foetus; p = expulsion of placenta. (Reproduced with permission from Fuchs, 1969 b.)

Poblete, 1970). Thus, in both species, parturition takes place at a time when uterine sensitivity to oxytocin is maximal.

The duration of labour in the rat is much longer than in the rabbit; 4 to 4·5 h, on average. The expulsion of a whole litter lasts about 2 h, on average. The expulsive stage is always preceded by increased uterine activity of 2 to 3 h duration (Fig. 7). Uterine activity remains of high intensity throughout the delivery of the litter but disappears soon after the last foetus and placenta has been cast. While the uterine contractions during labour do not resemble those produced by a single injection of oxytocin, contractions elicited by infusions of oxytocin on day 22 could not be distinguished from those recorded during spontaneous labour (Fig. 8). Delivery occurred during oxytocin infusions in 75 per cent of the rats while in rats receiving saline infusions less than 20 per cent delivered. The duration and outcome of oxytocin-induced labour did not differ from spontaneous parturition (Fuchs & Poblete, 1970). By contrast, the infusion of other agents which act on the uterus such as noradrenaline or prostaglandins E_1, E_2 or $F_{2\alpha}$ did not result in labour, except in a few instances.

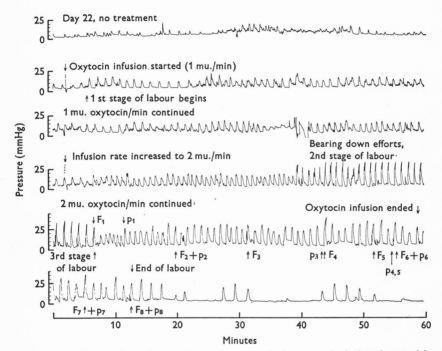

Fig. 8. Intrauterine pressure changes in a pregnant rat during oxytocin-induced parturition on day 22. F = expulsion of foetus; p = expulsion of placenta.

Furthermore, noradrenaline as well as prostaglandins were found to be more effective in stimulating uterine contractions on day 21 than on day 22. This is contrary to the observations with oxytocin (Figs. 9a and b). Vasopressin is very ineffective in stimulating uterine contractions in rats, and the rats infused with this hormone did not go into labour during the infusion. On the other hand, when both neurohypophysial hormones were infused together, all rats delivered during the infusion. Figure 10 summarises the results on induction of labour (Fuchs, 1972a). In spite of their greater potency on day 21, neither noradrenaline nor the prostaglandins induced premature delivery on day 21.

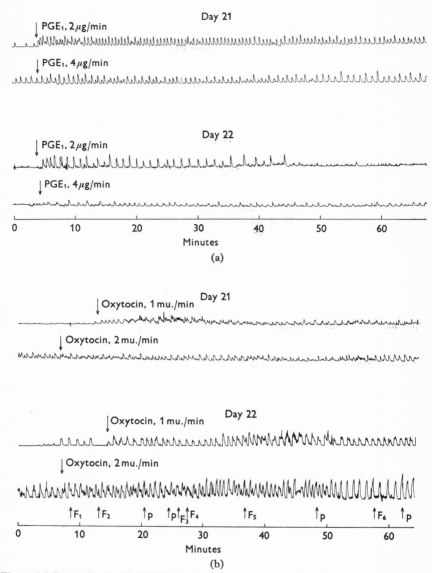

Fig. 9. (a) Prostaglandin E_1 (PGE$_1$)-induced uterine contractions in a rat on days 21 and 22 of pregnancy. No delivery occurred during the infusion. (b) Oxytocin-induced uterine contractions in a rat on days 21 and 22 of pregnancy. During the infusion the whole litter was expelled. F = expulsion of foetus; p = expulsion of placenta. (Printed with permission from Fuchs, 1972 a).

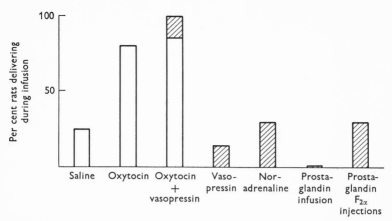

Fig. 10. Incidence of parturition in pregnant rats during a 4 h infusion of various substances at term. Prostaglandin $F_{2\alpha}$ injections: 20 or 50 μg, i.m. every 15 to 20 min for 4 h.
For other dosages see Fuchs (1972 a); 6 to 8 rats in each group. White bars = complete
delivery; shaded bars = partial or arrested delivery.

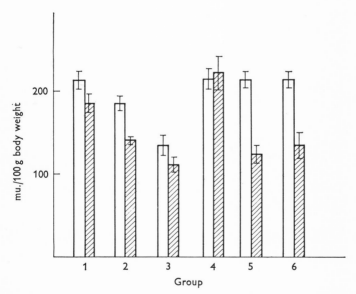

Fig. 11. Pituitary content of oxytocin (white bars) and vasopressin (shaded bars) in rats
before, during and after spontaneous or induced delivery, expressed as mu./100 g body
weight. Group 1 = saline-infused controls on day 21 and 22, not in labour (12 rats).
Group 2 = saline-infused controls killed immediately after delivery of first young.
Group 3 = saline-infused controls killed immediately after whole litter expelled. Group
4 = Oxytocin-infused rats killed on day 21, not in labour (7 rats). Group 5 = oxytocin-
infused rats killed immediately after delivery completed. Group 6 = oxytocin plus
vasopressin-infused rats killed after delivery of young was completed. The vertical lines
indicate ± s.e.m. Eight rats in each group unless otherwise indicated. (Printed with
permission from Fuchs & Saito, 1971.)

In an attempt to demonstrate neurohypophysial activation during parturition, we measured the pituitary content of oxytocin and vasopressin in rats before, during and immediately after delivery (Fuchs & Saito, 1971). The results (Fig. 11) clearly show that a considerable depletion of the pituitary content of both neurohypophysial peptides occurs during spontaneous labour. During parturition induced by oxytocin or by oxytocin and vasopressin together, the neurohypophysial oxytocin content

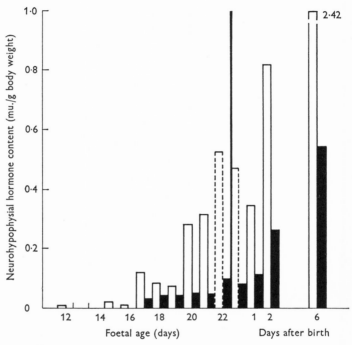

Fig. 12. Mean hormone content of the neurohypophysis in foetal and neonatal rats determined by radioimmunoassay. The white bars represent values for arginine vaso-pressin and the black bars oxytocin. The bars outlined by a broken line represent values from four litters in which the pituitaries were removed from different foetuses of the litter immediately before or immediately after delivery. (From a figure supplied by Dr M. L. Forsling based on unpublished results.)

did not fall whereas the vasopressin content decreased as in spontaneous labour. In rats giving birth spontaneously the oxytocin content was already depleted at the expulsion of the first foetus although the greatest decrease occurred during the expulsion of the entire litter. These observations suggest that the release of vasopressin is a consequence of labour whereas oxytocin is released for the stimulation, and not as the consequence, of labour contractions. Our results were in good agreement with those of

Achet, Chauvet & Olivry (1956). It is also of importance that the decrease in pituitary oxytocin content was of the same magnitude as the amount of oxytocin needed to induce labour in rats at term. On the other hand, the hormone content of the neurohypophysis of foetal rats does not decrease during delivery (M. L. Forsling, personal communication) (Fig. 12).

We have also used ethanol infusions in the rat to block neurohypophysial release. Unfortunately, very high doses of ethanol are needed to inhibit the release of oxytocin in this species. Experiments in lactating rats showed that over 4 g ethanol/kg had to be administered to the female in order to block oxytocin release in response to the suckling stimulus (Fuchs, 1969 a). The inhibition by ethanol was found to be dose dependent; about 50 per cent inhibition was observed with 2·5 g/kg. In addition,

Table 1. *Influence of ethanol on the onset and duration of labour in pregnant rats*

Dose of ethanol* (g/kg/h)	No. of rats	Gestation length at expulsion of first foetus (h)	Duration of delivery, F_1 to F_n (h)
0 (controls)	4	524·5 ± 2·6	2·1 ± 0·3
0·75–1·0	7	523·2 ± 2·3	2·1 ± 0·6
1·5–1·9	4	527·7 ± 3·3	3·7 ± 2·2
2·0–2·4	8	537·5 ± 3·1	4·2–∞

* This dose was given for 2 h in the beginning of the experiment and was followed by a maintenance dose varying from 200 to 400 mg/kg body weight.

ethanol is also metabolised much more rapidly in rats than in human beings (at a rate of 350 to 400 mg/kg/h in rats versus 100 mg/kg/h in man). It is, therefore, very difficult to maintain sufficiently high concentrations for a prolonged period of time without getting other undesirable effects related to the high ethanol dose, such as respiratory depression, metabolic acidosis, etc.

A series of experiments was nevertheless carried out to see if ethanol treatment had any effect at all in parturient rats. Different doses of intravenously infused ethanol were used in these studies. Table 1 gives the results. When the initial dose was over 4 g/kg (given over 2 h) a moderate but significant prolongation of gestation occurred. Labour was difficult and prolonged. On the other hand, ethanol in doses from 1 to 2 g/kg had no effect on the onset or duration of labour, as could be expected on the basis of the experiments in lactating rats.

All our results support the contention that oxytocin plays an essential part in the process of parturition in rats as well as in rabbits. Conflicting results have been reported by Gale & McCann (1961). They studied the effect of hypothalamic lesions on gestation and parturition in rats. In many rats the lesions resulted in severe diabetes insipidus, but a concomitant impairment of parturition was observed in only 33 per cent while the remainder delivered a living litter at term. A failure of lactation was observed in all rats. This was attributed to the absence of a milk-ejection reflex and hence of oxytocin and it was concluded that oxytocin was not essential for parturition. In a subsequent publication, the authors demonstrated that the lesions also caused a severe impairment of milk secretion (Gale, Taleisnik, Friedman & McCann, 1961). Replacement therapy with oxytocin alone did not significantly improve the milk yield. Thus, a separate deficiency of oxytocin and anterior pituitary hormones could not be demonstrated in these rats. Although a deficiency of vasopressin was produced, as shown by the increased water intake, oxytocin secretion could well have been relatively unimpaired. It is by now well established that the two peptides are synthesised and secreted by separate neurones in the hypothalamus.

It has recently been suggested that foetal rather than maternal neurohypophysial hormones are acting on the myometrium during labour (Chard, Hudson, Edwards & Boyd, 1971). However, in both rats and rabbits, the myometrial stimulant must be of maternal origin. This is shown by the drastic fall in the content of the maternal posterior pituitary hormones after delivery, while that in the foetal neurohypophysis did not fall but remained constant (M. L. Forsling, 1972, personal communication). The systemic nature of the myometrial stimulus is further demonstrated by the contractility pattern of the myometrial transplant and of the non-pregnant horn of the unilaterally pregnant uterus. If prostaglandins of uterine origin play a part in uterine activation during labour, their role must for the same reason be only a minor one, perhaps to facilitate and potentiate oxytocin-induced contractions, as suggested by several authors.

However, oxytocin is only one of the many factors responsible for the onset of labour, as clearly demonstrated by the fact that oxytocin cannot induce labour more than 6–8 h before term in rats, and about 24 h before expected delivery in rabbits. Nevertheless, in the rat, oxytocin can induce considerable uterine activity long before term, but these contractions do not bring about delivery of the litter before day 22. Uterine contractions as such are not, therefore, the only determining factor for the onset of labour. It should also be kept in mind that the intrauterine pressure changes within one segment of the uterus of a polytocous animal do not necessarily represent the changes in the uterine cavity as a whole. To gain more information on the uterine function of these multiparous animals, two intrauterine balloons were inserted into the same horn, but separated by two or more conceptuses, permitting simultaneous recording of intrauterine pressure changes from different locations (Fuchs & Poblete, 1970). The results showed that oxytocin stimulated all parts of the uterus through-out gestation. Before term contractions appeared to be initiated at random at both ends of the uterine horn, and the pressure waves remained localised

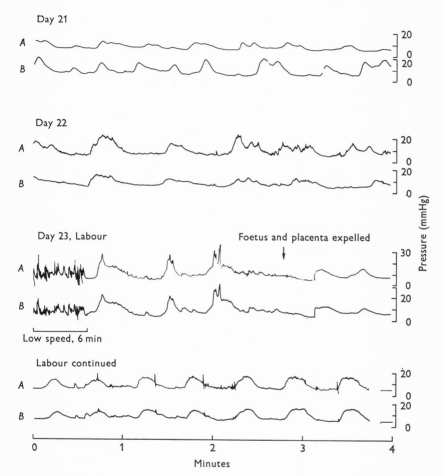

Fig. 13. Oxytocin-induced contractions of a pregnant rat uterus recorded simultaneously from two different segments of the same horn using a high paper speed. Recordings were made in late pregnancy (days 21–22), at term and during labour. A = ovarian end of the uterine horn; B = cervical end.

and segmental and were not conducted along the whole length of the uterus. Figure 13 illustrates these findings. As term approached the distal end of the uterus became more synchronous with the ovarian end. During labour, activity at the ovarian end seemed to predominate. Uterine pressure waves appeared to arise here and were rapidly conducted to the distal end which now was contracting in almost complete synchrony with the ovarian end. Spontaneous contractions of the rat uterus had the same characteristics as those induced by oxytocin (Fig. 14). Both ends contracted independently of each other during gestation, but were completely synchronous during labour. Synchronous contractions were recorded in each instance of oxytocin-induced as well as spontaneous parturition, and were never

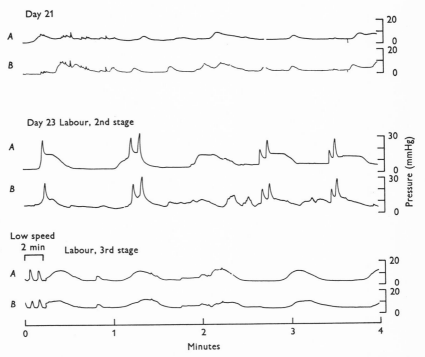

Fig. 14. Spontaneous contraction of a pregnant rat uterus recorded simultaneously from two separate segments of the same horn in late gestation and during labour. A = ovarian end of the uterine horn; B = cervical end.

observed before day 22. Thus, synchronous and well co-ordinated uterine activity appears to be a prerequisite for successful labour.

Oxytocin could not bring about this kind of activity more than 4–8 h before spontaneous parturition. Prostaglandins, on the other hand, did not produce such activity even at term and in fact appeared to impair the propagation of pressure waves. In concentrations that were high enough to elicit effective contractions at the ovarian end of the uterus, the cervical end contracted poorly, if at all. This is illustrated in Fig. 15 which shows uterine contractions in a rat in which induction of labour was attempted with intermittent injections of prostaglandin $F_{2\alpha}$ ($PGF_{2\alpha}$). In the same rat, oxytocin induced well co-ordinated, synchronous activity.

What are the factors responsible for this change in the mechanical properties of the myometrial muscle cells? According to observations by West & Landa (1956) and by Daniel (1960), stretch is a factor that can mediate interfibre spread of excitation in the rat uterus. The condition of the pregnant uterus just before parturition certainly favours such stretch-

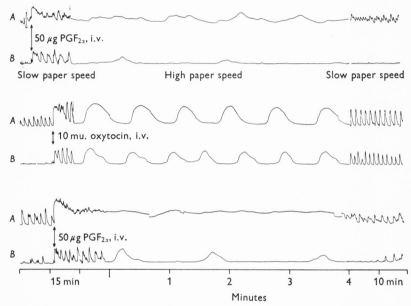

Fig. 15. Prostaglandin $F_{2\alpha}$ ($PGF_{2\alpha}$) and oxytocin-induced uterine contractions in a pregnant rat at term (day 22), recorded simultaneously from two balloons placed at the ovarian end (A) and the cervical end (B) of the uterine horn.

induced propagation of contractions. The uterus of the pregnant rat stops growing during the last few days of gestation, while the conceptuses more than double their weight from day 20 to day 22. At term, the conceptuses lie close to each other and occupy the whole of the uterine cavity so that the uterine horns assume a smooth cylindrical shape. However, when we examined the nonpregnant horn of unilaterally pregnant rats with the same technique, we found a change similar to that seen in the pregnant horn (Fig. 16), namely, a change from asynchronous activity during gestation to synchronous activity at term, thus indicating that stretch can only be a contributing factor and that the real cause is systemic in nature.

This view is further strengthened by observations made in rats in which pregnancy was artificially prolonged by means of progesterone injections.

In such animals foetal growth continues for about 2 days, while uterine weight does not increase. As a result the myometrium is progressively stretched beyond the degree seen at normal term. Recordings of intra-uterine pressure in such rats showed that uterine activity increased during

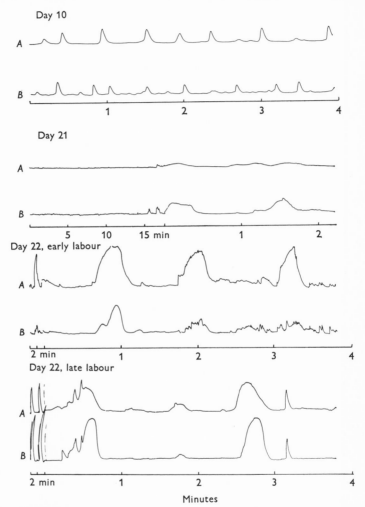

Fig. 16. Simultaneous recording of intrauterine pressure from both ends of the non-pregnant horn of a unilaterally pregnant rat. A = ovarian end; B = cervical end of the uterine horn.

the progesterone treatment and gradually became more intense than that seen during normal parturition, but the rats did not go into labour (Fig. 17). In spite of the marked stretching of the uterus under these conditions, the uterine activity in the two ends remained asynchronous.

Although the uterus responded to oxytocin with contractions of increased intensity, their synchrony was not improved by oxytocin. In spite of the intense and sustained uterine activity the cervix remained firmly closed.

Progesterone, therefore, appeared to impair the propagation of contractile activity, as suggested by Csapo (1956a) and Kuriyama & Csapo (1961). Quite clearly, progesterone does not block uterine activity as such, nor does it hinder the mechanism of excitation, at least in the rat uterus. Others have failed to demonstrate a blocking action of progesterone on the

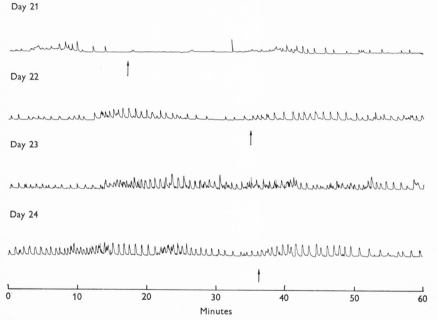

Fig. 17. Recordings of intrauterine pressure changes in a pregnant rat treated with progesterone (5 mg/day, i.m.) starting on day 21. Parturition did not occur in spite of strong contractions elicited by daily infusions of 1 mu. oxytocin/min (arrows).

electrical and mechanical properties of the rat uterus, and it has been suggested that lack of propagation can better be explained by withdrawal of oestrogens (Melton & Saldivar, 1965; Saldivar & Melton, 1966).

Observations in ovariectomised pregnant rats corroborate this concept. After the removal of the ovaries, an increase in uterine activity was evident within a few hours and 24 or 48 h later the activity was usually very intense. Simultaneous recordings from two separate sites showed that activity in these rats remained asynchronous in spite of the intensity of the contractions. Oxytocin further increased the intensity of the uterine activity but failed to cause a synchronisation of the contractions of the two ends of the same uterine horn.

An indication of the importance of such synchronous activity for successful parturition is the fact that the rat rarely aborts living offspring. If ovariectomy is performed early, before day 13, all the conceptuses are resorbed. After ovariectomy between days 14 and 17, many are resorbed but some foetuses may survive *in utero* until day 22 but labour does not occur or is very difficult and placentas are invariably retained. After ovariectomy on days 18 or 19, the foetuses are usually expelled within about 50 to 60 h but in an abnormal fashion, one by one, at long intervals, and are either dead at delivery or die very soon afterwards, and are usually eaten by the mother. Ovariectomy on days 20 or 21 generally leads to postponement of delivery to day 23 or 24. Delivery is always very prolonged and difficult. However, regardless of the day of ovariectomy, strong uterine activity always develops within 24 h.

When rats ovariectomised on day 20 are given oestrogen injections on day 21, the expulsion of the litter occurs within the normal time limits (Acker, 1969), and our findings were the same in rats ovariectomised on day 21. Recordings of uterine activity in these rats demonstrated that the pattern of delivery was not restored to normal; it was, nevertheless, significantly improved as compared with untreated ovariectomised rats. It is possible that the degree of deviation from the normal pattern of delivery depends on the dosage of oestrogens. The possibility that other ovarian hormones, such as relaxin, are also of importance for the normal process of parturition is indicated by the studies of Kroc, Steinetz & Beach (1959) and should be re-examined.

The effect of $PGF_{2\alpha}$ on pregnant rats on days 18 to 20 further emphasises the importance of a maintained or probably increased oestrogen secretion together with a withdrawal of progesterone for the normal course of labour. While all three prostaglandins failed as oxytocics and did not induce labour at term, $PGF_{2\alpha}$ in quite low doses resulted in premature but delayed delivery, when infused on days 18 or 20. Labour occurred after an interval of 41·0 and 33·5 h, respectively and was essentially normal with respect to uterine activity, duration and outcome. It is probably the luteolytic effect of $PGF_{2\alpha}$ that is responsible for this action. Progesterone levels were found to decline to 20 per cent of the initial values during an infusion of 6 h duration, and non-pregnant values were observed 48 h later (Fuchs, Mok & Sundaram, 1972). Oestrogen levels were not determined, but ovulation had occurred within 24 to 36 h of the premature delivery in all $PGF_{2\alpha}$-treated rats, indicating an accelerated follicular maturation and presumably also oestrogen secretion.

In conclusion, parturition in rats and rabbits occurs as a result of a withdrawal of progesterone with simultaneous or subsequent increase in oestrogen production which leads to sensitisation of the uterus to oxytocin stimulation and to a change in the excitatory properties of the myometrium, thus making synchronous uterine activity possible.

This change in hormone balance at term probably also affects the hypothalamo-neurohypophysial neurosecretory system. According to Roberts & Share (1969), progesterone diminishes, and oestrogen increases oxytocin release in response to vaginal stimulation. On the basis of the available evidence, there is reason to believe that oestrogens might also increase the biosynthesis of oxytocin in the hypothalamic oxytocinergic neurones.

Neurohypophysial activation with oxytocin release is the final event in the termination of gestation and causes the expulsion of the young. It remains to be discovered what brings about the initial event, namely, the demise of the corpora lutea of pregnancy, and what is the stimulus that results in the release of oxytocin as the final event in the process of parturition.

REFERENCES

ACHET, R., CHAUVET, J. & OLIVRY, G. (1956). Sur l'existence éventuelle d'une hormone unique neurohypophysaire. *Biochim. biophys. Acta*, **22**, 428–433.

ACKER, G. (1969). A propos du déterminisme hormonal du mise bas chez la ratte; rôle de l'oestradiol. *C.r. hebd. Séanc. Acad. Sci., Paris*, **268**, 2196–2199.

BELING, C., FUCHS, A.-R. & FLORENCIO, C. (1973). In preparation.

BENGTSSON, L. PH. (1957). The endocrine control of myometrial contractility in the uterus of the pregnant rabbit. *Am. J. Obstet. Gynec.* **74**, 480–499.

CHARD, T., HUDSON, C. N., EDWARDS, C. R. W. & BOYD, N. R. H. (1971). Release of oxytocin and vasopressin by the human foetus during labour. *Nature, Lond.* **234**, 352–353.

COUTINHO, E. M. & DE MATTOS, C. E. R. (1968). Effects of estrogens on the non-atrophic estrogen-deficient rabbit uterus. *Endocrinology*, **83**, 422–432.

CSAPO, A. (1956a). The mechanism of effect of ovarian steroids. *Recent Prog. Horm. Res.* **12**, 405–427.

CSAPO, A. (1956b). Progesterone 'block'. *Am. J. Anat.* **98**, 273–291.

CSAPO, A. I. & LLOYD-JACOB, M. A. (1962). Placenta, uterine volume and the control of the pregnant uterus in rabbits. *Am. J. Obstet. Gynec.* **83**, 1073–1082.

CSAPO, A. I. & WIEST, W. G. (1969). An examination of the quantitative relationship between progesterone and the maintenance of pregnancy. *Endocrinology*, **85**, 735–746.

DANIEL, E. E. (1960). The activation of various types of uterine muscle during stretch-induced contraction. *Can. J. Physiol.* **38**, 1327–1362.

FUCHS, A.-R. (1964a). Oxytocin and the onset of labour in rabbits. *J. Endocr.* **30**, 217–227.

FUCHS, A.-R. (1964b). The role of oxytocin in the initiation of labour. In *Proc. 2nd International Congress of Endocrinology, London*, ed. S. Taylor, pp. 753–758, Int. Congr. Series no. 83. Amsterdam: Excerpta Medica Foundation.

FUCHS, A.-R. (1966). The inhibitory effect of ethanol on the release of oxytocin during parturition in the rabbit. *J. Endocr.* **35**, 125–134.

FUCHS, A.-R. (1969a). Ethanol and the inhibition of oxytocin release in lactating rats. *Acta endocr., Copenh.* **62**, 546–554.

FUCHS, A.-R. (1969b). Uterine activity in late pregnancy and during parturition in the rat. *Biology of Reproduction*, **1**, 344–353.

FUCHS, A.-R. (1972a). Prostaglandin effects on rat pregnancy. I. Failure of induction of labor. *Fert. Steril.* **23**, 410–416.

FUCHS, A.-R. (1972b). Uterine activity during and after mating in the rabbit. *Fert. Steril.* **23**, 915–923.

FUCHS, A.-R. & FUCHS, F. (1960). The effect of oxytocic substances upon the rabbit uterus *in situ*. *Acta physiol. scand.* **49**, 103–113.

FUCHS, A.-R., MOK, E. & SUNDARAM, K. (1972). Prostaglandin effects on luteal function in rats. In *Abstracts of the Proceedings of the 4th International Congress of Endocrinology, Washington, D.C.*, Abstr. no. 474, p. 188. Int. Congr. Series no. 256. Amsterdam: Excerpta Medica Foundation.

FUCHS, A.-R. & POBLETE, V. F. Jr (1970). Oxytocin and uterine function in pregnant and parturient rats. *Biol. Reprod.* **2**, 387–400.

FUCHS, A.-R. & SAITO, S. (1971). Pituitary oxytocin and vasopressin content of pregnant rats before, during and after parturition. *Endocrinology*, **88**, 574–578.

FUCHS, A.-R. & VIEIRA LOPES, A. C. (1970). Failure of 1-L-penicillamine-oxytocin and 1-deaminopenicillamine-oxytocin to inhibit the uterine effect of oxytocin in rabbit *in vivo*. *Endocrinology*, **86**, 71–76.

GALE, C. C. & MCCANN, S. M. (1961). Hypothalamic control of pituitary gonadotropins. *J. Endocr.* **22**, 107–117.

GALE, C. C., TALEISNIK, S., FRIEDMAN, H. M. & MCCANN, S. M. (1961). Hormonal basis for impairment in milk synthesis and milk ejection following hypothalamic lesions. *J. Endocr.* **23**, 303–316.

HEROD, L., MORIER, J., LAWLESS, J. & LANMAN, J. T. (1972). Improved viability of offspring with estrogen supplementation of ovariectomized rabbits maintained on progesterone. *J. Reprod. Fert.* **29**, 137–140.

HILLIARD, J., SPIES, H. G. & SAWYER, C. H. (1968). Cholesterol storage and progestin secretion during pregnancy and pseudopregnancy in the rabbit. *Endocrinology*, **82**, 157–165.

KEYES, P. L. & ARMSTRONG, D. T. (1968). Endocrine role of follicles in the regulation of corpus luteum function in the rabbit. *Endocrinology*, **83**, 509–515.

KEYES, P. L. & NALBANDOV, A. V. (1967). Maintenance and function of corpora lutea in rabbits depend on estrogen. *Endocrinology*, **80**, 938–946.

KNAUS, H. H. (1930a). Zur Physiologie des Corpus Luteum, II. *Arch. exp. Path. Pharmak.* **140**, 181–190.

KNAUS, H. H. (1930b). Zur Physiologie des Corpus Luteum, III. *Arch. exp. Path. Pharmak.* **141**, 374–394.

KNAUS, H. H. (1930c). Zur Physiologie des Corpus Luteum, IV. *Arch. exp. Path. Pharmak.* **141**, 395–403.

KROC, R. L., STEINETZ, B. G. & BEACH, V. L. (1959). The effects of estrogens, progestagens and relaxin in pregnant and nonpregnant laboratory rodents. *Ann. N.Y. Acad. Sci.* **75**, 942–980.

KURIYAMA, H. & CSAPO, A. (1961). A study of the parturient uterus with the microelectrode technique. *Endocrinology*, **68**, 1010–1025.

MAKEPEACE, A. W., CORNER, G. W. & ALLEN, W. M. (1936). The effect of progestin on the in vitro response of the rabbit's uterus to pituitrin. *Am. J. Physiol.* **115**, 376–385.

MELTON, C. E. & SALDIVAR, J. T. JR (1965). Estrus cycle and electrical activity of rat myometrium. *Life Sci.* **4**, 594–602.

REYNOLDS, S. R. M. (1949). *Physiology of the Uterus*, 2nd edn. New York: Paul Hoeber.

ROBERTS, J. S. & SHARE, L. (1969). Effects of progesterone and estrogen on blood levels of oxytocin during vaginal distension. *Endocrinology*, **84**, 1076–1081.

ROBSON, J. M. (1937). Maintenance by oestrin of the luteal function in hypophysectomized rabbits. *J. Physiol., Lond.* **90**, 125–144.

SALDIVAR, J. T. & MELTON, C. E. (1966). Effects *in vivo* and *in vitro* of sex steroids on rat myometrium. *Am. J. Physiol.* **221**, 835–843.

SCHOFIELD, B. M. (1957). The hormonal control of myometrial function during pregnancy. *J. Physiol., Lond.* **138**, 1–10.

TAKEDA, H. (1965). Generation and propagation of uterine activity *in situ*. *Fert. Steril.* **16**, 113–119.

WAGNER, G. (1969). Electrical and mechanical activity in myometrial grafts in the rabbit. *Acta physiol. scand.* **76**, 1A–2A.

WEST, T. C. & LANDA, J. (1956). Transmembrane potentials and contractility in the pregnant rat uterus. *Am. J. Physiol.* **187**, 333–337.

WESTMAN, A. (1934). Untersuchungen über die Abhängigkeit der Funktion des Corpus Luteum von der Ovarialfollikeln und über die Bildungsstette der Hormone im Ovarium. *Arch. Gynaek.* **158**, 476–504.

INDEX